International Vocational Education Bilingual Textbook Series

国际化职业教育双语系列教材

Electrical Control and PLC Application
电气控制与 PLC 应用

Wang Zhixue
王治学　　主　编

Xiao Jun　　Tang Jiying　　Zhang Tao
肖　军　　唐继英　　张　涛　　副主编

Beijing
Metallurgical Industry Press
2020

内 容 提 要

本书共分四个项目：项目1主要介绍电气控制与PLC基础知识，包含3个任务，内容主要涉及电气控制与PLC基础知识、PLC基础知识和编程软件的使用；项目2主要介绍三相异步电动机的瞬时启停控制，包含2个任务，内容主要涉及三相异步电动机的点动自锁控制和三相异步电动机的正反转控制；项目3主要介绍三相异步电动机的延时控制，包含3个任务，内容主要涉及电动机的延时启动主电路及控制电路的设计与实现，电动机的延时停止主电路及控制电路的设计与实现，以及电动机的循环启停主电路及控制电路的设计与实现；项目4主要介绍三相异步电动机的变频器控制，包含2个任务，内容主要涉及利用变频器面板控制三相异步电动机的运行和利用PLC和变频器控制三相异步电动机运行。

本书可作为职业院校机电和电气类专业的国际化教学用书，也可作为机电、电气行业工程技术人员的国际化培训教材。

图书在版编目(CIP)数据

电气控制与PLC应用=Electrical Control and PLC Application/王治学主编. —北京：冶金工业出版社，2020.9
国际化职业教育双语系列教材
ISBN 978-7-5024-8533-7

Ⅰ.①电⋯ Ⅱ.①王⋯ Ⅲ.①电气控制—高等职业教育—双语教学—教材—汉、英 ②PLC技术—高等职业教育—双语教学—教材—汉、英 Ⅳ.①TM571.2 ②TM571.6

中国版本图书馆CIP数据核字(2020)第209612号

出 版 人 苏长永
地　　址 北京市东城区嵩祝院北巷39号　邮编 100009　电话 (010)64027926
网　　址 www.cnmip.com.cn　电子信箱 yjcbs@cnmip.com.cn
责任编辑 俞跃春 刘林烨　美术编辑 郑小利　版式设计 孙跃红 禹 蕊
责任校对 李 娜 责任印制 李玉山

ISBN 978-7-5024-8533-7
冶金工业出版社出版发行；各地新华书店经销；三河市双峰印刷装订有限公司印刷
2020年9月第1版，2020年9月第1次印刷
787mm×1092mm 1/16；19印张；450千字；279页
58.00元

冶金工业出版社　投稿电话　(010)64027932　投稿信箱　tougao@cnmip.com.cn
冶金工业出版社营销中心　电话　(010)64044283　传真　(010)64027893
冶金工业出版社天猫旗舰店　yjgycbs.tmall.com

(本书如有印装质量问题，本社营销中心负责退换)

Editorial Board of International Vocational Education Bilingual Textbook Series

Director Kong Weijun (Party Secretary and Dean of Tianjin Polytechnic College)

Deputy Director Zhang Zhigang (Chairman of Tiantang Group, Sino-Uganda Mbale Industrial Park)

Committee Members Li Guiyun, Li Wenchao, Zhao Zhichao, Liu Jie, Zhang Xiufang, Tan Qibing, Liang Guoyong, Zhang Tao, Li Meihong, Lin Lei, Ge Huijie, Wang Zhixue, Wang Xiaoxia, Li Rui, Yu Wansong, Wang Lei, Gong Na, Li Xiujuan, Zhang Zhichao, Yue Gang, Xuan Jie, Liang Luan, Chen Hong, Jia Yanlu, Chen Baoling

国际化职业教育双语系列教材编委会

主　任　孔维军（天津工业职业学院党委书记、院长）

副主任　张志刚（中乌姆巴莱工业园天唐集团董事长）

委　员　李桂云　李文潮　赵志超　刘　洁　张秀芳

　　　　　谭起兵　梁国勇　张　涛　李梅红　林　磊

　　　　　葛慧杰　王治学　王晓霞　李　蕊　于万松

　　　　　王　磊　宫　娜　李秀娟　张志超　岳　刚

　　　　　玄　洁　梁　娈　陈　红　贾燕璐　陈宝玲

Foreword

With the proposal of the 'Belt and Road Initiative', the Ministry of Education of China issued *Promoting Education Action for Building the Belt and Road Initiative* in 2016, proposing cooperation in education, including 'cooperation in human resources training'. At the Forum on China-Africa Cooperation (FOCAC) in 2018, President Xi proposed to focus on the implementation of the 'Eight Actions', which put forward the plan to establish 10 Luban Workshops to provide skills training to African youth. Draw lessons from foreign advanced experience of vocational education mode, China's vocational education has continuously explored and formed the new mode of vocational education with Chinese characteristics. Tianjin, as a demonstration zone for reform and innovation of modern vocational education in China, has started the construction of 'Luban Workshop' along the 'Belt and Road Initiative', to export high-quality vocational education achievements.

The compilation of these series of textbooks is in response to the times and it's also the beginning of Tianjin Polytechnic College to explore the internationalization of higher vocational education. It's a new model of vocational education internationalization by Tianjin, response to the 'Belt and Road Initiative' and the 'Going Out' of Chinese enterprises. Tianjin Polytechnic College and Uganda Technical College-Elgon reached a cooperation intention to establish the Luban Workshop to carry out vocational education cooperation on mechatronics technology and ferrous metallurgy technology major in 2019. The establishment of Luban Workshop is conducive to strengthen the cooperation between China and Uganda in vocational education, promote the export of high-quality higher vocational education resources, and serve Chinese enterprises in Uganda and Ugandan local enterprises. Exploring and standardizing the overseas operation of Chinese colleges, the expansion of international influences of China's higher vocational education is also one of the purposes.

The construction of 'Luban Workshop' in Uganda is mainly based on the EPIP (Engineering, Practice, Innovation, Project) project, and is committed to cultivating high-quality talents with innovative spirit, creative ability and entrepreneurial spirit. To meet the learning needs of local teachers and students accurately, the compilation of these international vocational skills bilingual textbooks is based on the talent demand of Uganda and the specialty and characteristics of Tianjin Polytechnic.

These textbooks are supporting teaching material, referring to Chinese national professional standards and developing international professional teaching standards. The internationalization of the curriculums takes into account the technical skills and cognitive characteristics of local students, to promote students' communication and learning ability. At the same time, these textbooks focus on the enhancement of vocational ability, rely on professional standards, and integrate the teaching concept of equal emphasis on skills and quality. These textbooks also adopted project-based, modular, task-driven teaching model and followed the requirements of enterprise posts for employees.

In the process of writing the series of textbooks, Wang Xiaoxia, Li Rui, Wang Zhixue, Ge Huijie, Yu Wansong, Wang Lei, Li Xiujuan, Gong Na, Zhang Zhichao, Jia Yanlu, Chen Baoling and other chief teachers, professional teams, English teaching and research office have made great efforts, receiving strong support from leaders of Tianjin Polytechnic College. During the compilation, the series of textbooks referred to a large number of research findings of scholars in the field, and we would like to thank them for their contributions.

Finally, we sincerely hope that the series of textbooks can contribute to the internationalization of China's higher vocational education, especially to the development of higher vocational education in Africa.

Principal of Tianjin Polytechnic College　Kong Weijun
May, 2020

序

随着"一带一路"倡议的提出，2016 年中华人民共和国教育部发布了《推进共建"一带一路"教育行动》，提出了包括"开展人才培养培训合作"在内的教育合作。2018 年习近平主席在中非合作论坛上提出，要重点实施"八大行动"，明确要求在非洲设立 10 个鲁班工坊，向非洲青年提供技能培训。中国职业教育在吸收和借鉴发达国家先进职教发展模式的基础上，不断探索和形成了中国特色职业教育办学模式。天津市作为中国现代职业教育改革创新示范区，开启了"鲁班工坊"建设工作，在"一带一路"沿线国家搭建"鲁班工坊"平台，致力于把优秀职业教育成果输出国门与世界分享。

本系列教材的编写，契合时代大背景，是天津工业职业学院探索高职教育国际化的开端。"鲁班工坊"是由天津率先探索和构建的一种职业教育国际化发展新模式，是响应国家"一带一路"倡议和中国企业"走出去"，创建职业教育国际合作交流的新窗口。2019 年天津工业职业学院与乌干达埃尔贡技术学院达成合作意向，共同建立"鲁班工坊"，就机电一体化技术专业、黑色冶金技术专业开展职业教育合作。此举旨在加强中乌职业教育交流与合作，推动中国优质高等职业教育资源"走出去"，服务在乌中资企业和乌干达当地企业，探索和规范我国职业院校"鲁班工坊"建设和境外办学，扩大中国高等职业教育的国际影响力。

中乌"鲁班工坊"的建设主要以工程实践创新项目（EPIP：Engineering, Practice, Innovation, Project）为载体，致力于培养具有创新精神、创造能力和创业精神的"三创"复合型高素质技能人才。国际化职业教育双语系列教材的编写，立足于乌干达人才需求和天津工业职业学院专业特色，是为了更好满足当地师生学习需求。

本系列教材采用中英双语相结合的方式，主要参照中国专业标准，开发国际化专业教学标准，课程内容国际化是在专业课程设置上，结合本地学生的技术能力水平与认知特点，合理设置双语教学环节，加强学生的学习与交流能力。同时，教材以提升职业能力为核心，以职业标准为依托，体现技能与质量并重的教学理念，主要采用项目化、模块化、任务驱动的教学模式，并结合企业岗位对员工的要求来撰写。

本系列教材在撰写过程中，王晓霞、李蕊、王治学、葛慧杰、于万松、王磊、李秀娟、宫娜、张志超、贾燕璐、陈宝玲等主编老师、专业团队、英语教研室付出了辛勤劳动，并得到了学院各级领导的大力支持，同时本系列教材借鉴和参考了业界有关学者的研究成果，在此一并致谢！

最后，衷心希望本系列教材能为我国高等职业教育国际化，尤其是高等职业教育走进非洲、支援非洲高等职业教育发展尽绵薄之力。

天津工业职业学院书记、院长　孔维军

2020 年 5 月

Preface

Tianjin Polytechnic College and Uganda Technical College-Elgon reached a cooperation intention to establish the Luban Workshop to carry out vocational education cooperation on mechatronics technology and ferrous metallurgy technology major in 2019. In order to strengthen the cooperation between China and Uganda in vocational education, the two colleges plan to compile a series of international vocational skills bilingual textbooks.

This book is one of the international vocational skills bilingual textbooks. The book was written by Wang Zhixue, Xiao Jun, Tang Jiying and Zhang Tao of Tianjin Polytechnic college. Wang Zhixue is responsible for the compilation and revision of the whole book, and the compilation of Project 1 and Project 2; Project 3 is compiled by Xiao Jun; Task 4.1 of Project 4 is compiled by Zhang Tao, and Task 4.2 is compiled by Tang Jiying. I would like to thank Zhao Xueqing, Wang Qingye and other teachers for their help in the preparation of the textbook.

The relevant materials and literature were referenced in the process of writing. Here express my gratitude to the author concerned.

Due to the limited level of the editor, there is something wrong in the book. I hope readers to criticize and correct.

The editor
February, 2020

前 言

2019年天津工业职业学院与乌干达埃尔贡技术学院达成合作意向，共同建立"鲁班工坊"，就机电一体化技术专业、黑色冶金技术专业开展职业教育合作，双方计划编撰国际化职业教育双语系列教材。

本书是国际化职业教育双语系列教材之一。本书编写工作由天津工业职业学院王治学、肖军、唐继英、张涛完成。王治学负责全书的统稿与修改，并编写项目1和项目2；项目3由肖军编写；项目4中的任务4.1由张涛编写，任务4.2由唐继英编写。在此对教材编写前期提供帮助的赵学庆、王青叶等老师表示感谢。

本书在编写过程中，参考了有关资料和文献，在此向有关作者表示感谢。

由于编者水平所限，书中不妥之处，希望读者批评指正。

编　者
2020年2月

Contents

Project 1 Basic Knowledge of Electrical Control and PLC 1

Task 1.1 Basic knowledge of Electrical Control and PLC 1
  1.1.1 Task Description 1
  1.1.2 Task Target 1
  1.1.3 Task-related Knowledge 1
  1.1.4 Task Implementation 11
  1.1.5 Task Evaluation 11
  1.1.6 Task Summary 12
  1.1.7 Task Development 13
Task 1.2 PLC Basics 13
  1.2.1 Task Description 13
  1.2.2 Task Target 13
  1.2.3 Task-related Knowledge 14
  1.2.4 Task Implementation 20
  1.2.5 Task Evaluation 20
  1.2.6 Task Summary 22
  1.2.7 Task Development 22
Task 1.3 Use of Programming Software 25
  1.3.1 Task Description 25
  1.3.2 Task Target 26
  1.3.3 Task-related Knowledge 26
  1.3.4 Task Implementation 32
  1.3.5 Task Evaluation 33
  1.3.6 Task Summary 34
  1.3.7 Task Development 35
  1.3.8 Project Summary 35

Project 2 Instantaneous Start Stop Control of Three-phase Asynchronous Motor 37

Task 2.1 Inching Self-locking Control of Three-phase Asynchronous Motor 37
  2.1.1 Task Description 37

2.1.2	Task Target	37
2.1.3	Task-related Knowledge	38
2.1.4	Task Implementation	47
2.1.5	Task Evaluation	50
2.1.6	Task Summary	51
2.1.7	Task Development	52

Task **2.2** Forward and Reverse Control of Three-phase Asynchronous Motor ······ 52

2.2.1	Task Description	52
2.2.2	Task Target	52
2.2.3	Task-related Knowledge	53
2.2.4	Task Implementation	57
2.2.5	Task Evaluation	62
2.2.6	Task Summary	63
2.2.7	Task Development	64
2.2.8	Project Summary	64

Project 3   Delay Control of Three-phase Asynchronous Motor ······ 66

Task **3.1** Wiring and Debugging of Main Circuit and Control Circuit of Delayed Start of Motor ······ 66

3.1.1	Task Description	66
3.1.2	Task Target	66
3.1.3	Task-related Knowledge	66
3.1.4	Task Implementation	70
3.1.5	Task Evaluation	74
3.1.6	Task Summary	75
3.1.7	Task Development	76

Task **3.2** Design and Implementation of Main Circuit and Control Circuit of Motor's Delayed Stop ······ 76

3.2.1	Task Description	76
3.2.2	Task Target	76
3.2.3	Task-related Knowledge	77
3.2.4	Task Implementation	82
3.2.5	Task Evaluation	88
3.2.6	Task Summary	89
3.2.7	Task Development	89

Task **3.3** Design and Implementation of Main Circuit and Control Circuit of Cyclic Start and Stop Control of Motor ······ 90

3.3.1	Task Description	90

3.3.2	Task Target	90
3.3.3	Task-related Knowledge	90
3.3.4	Task Implementation	96
3.3.5	Task Evaluation	102
3.3.6	Task Summary	103
3.3.7	Task Development	103
3.3.8	Project Summary	104

Project 4　Inverter Control of Three-phase Asynchronous Motor 108

Task 4.1　Using Inverter Panel to Control Operation of Three-phase Asynchronous Motor 108

4.1.1	Task Description	108
4.1.2	Task Target	108
4.1.3	Task-related Knowledge	109
4.1.4	Task Implementation	124
4.1.5	Task Evaluation	130
4.1.6	Task Summary	131
4.1.7	Task Development	131

Task 4.2　Using S7-1215C PLC and Inverter to Control Operation of Three-phase Asynchronous Motor 132

4.2.1	Task Description	132
4.2.2	Task Target	132
4.2.3	Task-related Knowledge	132
4.2.4	Task Implementation	135
4.2.5	Task Evaluation	146
4.2.6	Task Summary	148
4.2.7	Task Development	149
4.2.8	Project Summary	149

Reference 151

目 录

项目1 电气控制与PLC ... 152

任务1.1 电气控制与PLC基础知识 ... 152
- 1.1.1 任务描述 ... 152
- 1.1.2 任务目标 ... 152
- 1.1.3 任务相关知识 ... 152
- 1.1.4 任务实施 ... 160
- 1.1.5 任务评价 ... 160
- 1.1.6 任务小结 ... 161
- 1.1.7 任务拓展 ... 161

任务1.2 PLC基础知识 ... 161
- 1.2.1 任务描述 ... 161
- 1.2.2 任务目标 ... 162
- 1.2.3 任务相关知识 ... 162
- 1.2.4 任务实施 ... 167
- 1.2.5 任务评价 ... 167
- 1.2.6 任务小结 ... 168
- 1.2.7 任务拓展 ... 169

任务1.3 编程软件的使用 ... 172
- 1.3.1 任务描述 ... 172
- 1.3.2 任务目标 ... 172
- 1.3.3 任务相关知识 ... 172
- 1.3.4 任务实施 ... 179
- 1.3.5 任务评价 ... 180
- 1.3.6 任务小结 ... 181
- 1.3.7 任务拓展 ... 181
- 1.3.8 项目小结 ... 182

项目2 三相异步电动机的瞬时启停控制 ... 183

任务2.1 三相异步电动机的点动自锁控制 ... 183
- 2.1.1 任务描述 ... 183
- 2.1.2 任务目标 ... 183

2.1.3 任务相关知识 …… 183
2.1.4 任务实施 …… 190
2.1.5 任务评价 …… 193
2.1.6 任务小结 …… 194
2.1.7 任务拓展 …… 195

任务 2.2 三相异步电动机的正反转控制 …… 195
2.2.1 任务描述 …… 195
2.2.2 任务目标 …… 195
2.2.3 任务相关知识 …… 196
2.2.4 任务实施 …… 200
2.2.5 任务评价 …… 203
2.2.6 任务小结 …… 204
2.2.7 任务拓展 …… 205
2.2.8 项目小结 …… 205

项目 3 三相异步电动机的延时控制 …… 206

任务 3.1 电动机的延时启动主电路及控制电路的设计与实现 …… 206
3.1.1 任务描述 …… 206
3.1.2 任务目标 …… 206
3.1.3 任务相关知识 …… 206
3.1.4 任务实施 …… 209
3.1.5 任务评价 …… 213
3.1.6 任务小结 …… 213
3.1.7 任务拓展 …… 214

任务 3.2 电动机的延时停止主电路及控制电路的设计与实现 …… 214
3.2.1 任务描述 …… 214
3.2.2 任务目标 …… 214
3.2.3 任务相关知识 …… 214
3.2.4 任务实施 …… 219
3.2.5 任务评价 …… 224
3.2.6 任务小结 …… 225
3.2.7 任务拓展 …… 226

任务 3.3 电动机的循环启停主电路及控制电路的设计与实现 …… 226
3.3.1 任务描述 …… 226
3.3.2 任务目标 …… 226
3.3.3 任务相关知识 …… 226
3.3.4 任务实施 …… 232
3.3.5 任务评价 …… 236
3.3.6 任务小结 …… 237

3.3.7	任务拓展	238
3.3.8	项目小结	238

项目 4　三相异步电动机的变频器控制 　241

任务 4.1　利用变频器面板控制三相异步电动机的运行 　241

- 4.1.1　任务描述　241
- 4.1.2　任务目标　241
- 4.1.3　任务相关知识　241
- 4.1.4　任务实施　255
- 4.1.5　任务评价　260
- 4.1.6　任务小结　260
- 4.1.7　任务拓展　261

任务 4.2　利用 PLC 和变频器控制三相异步电动机运行 　261

- 4.2.1　任务描述　261
- 4.2.2　任务目标　261
- 4.2.3　任务相关知识　262
- 4.2.4　任务实施　264
- 4.2.5　任务评价　275
- 4.2.6　任务小结　276
- 4.2.7　任务拓展　277
- 4.2.8　项目小结　277

参考文献　279

Project 1　Basic Knowledge of Electrical Control and PLC

Electrical control and PLC technology is one of the professional technologies closely related to equipment manufacturing, which are widely used in various electrical equipment, CNC machine tools and production practices. This project is divided into three tasks: identification and drawing of electrical schematic diagram, learning of PLC basic knowledge, and programming software. Through the study of this project, students will master the basic knowledge of electrical schematic diagram. They are able to identify common low-voltage electrical components, to have a certain understanding of PLC, and to operate programming software skillfully. This is the basic link of this course, laying a solid foundation for the specific project task learning later.

Task 1.1　Basic Knowledge of Electrical Control and PLC

1.1.1　Task Description

This task is mainly to learn the basic knowledge of common electrical components and electrical schematic diagram. Through learning and observing the low-voltage electrical components, students should fill the corresponding information into the table. They also need draw the electrical schematic diagram, mark the wire number, and complete the simple wiring work. Finally, the process of task implementation is summarized and recorded in the corresponding table.

1.1.2　Task Target

(1) Master the principles and applications of common low-voltage electrical components.
(2) Master the basic drawing methods of electrical schematic diagram.
(3) Master the marking methods of wire number.
(4) Understand the basic knowledges of three-phase asynchronous motor.
(5) Train students' sense of safe operations and team work.

1.1.3　Task-related Knowledge

1.1.3.1　Low Voltage Circuit Breaker

The low voltage circuit breaker (once called automatic switch) is a kind of switch apparatus which can not only turn on and off the normal load current and overload current, but also turn on and off the short-circuit current. In addition to the control function, the low-voltage circuit breaker also has some protection functions, such as overload, short circuit, undervoltage and leakage protection. The capacity range of LV circuit breaker is very large. The minimum is 4A, and the maximum is 5000A. Low voltage circuit breakers are widely used in low-voltage distribution system, power control of various mechanical equipment and control and protection of power terminals. The actual drawing and electrical symbols are shown in Figures 1-1 and 1-2 respectively.

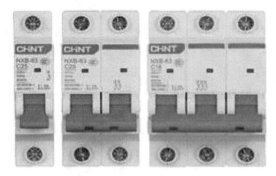

Figure 1-1 Physical drawing of
low voltage circuit breaker

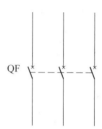

Figure 1-2 Electrical symbols of
low voltage circuit breaker

Contact

The contact mechanically connected with the opening and closing mechanism of the main circuit of the circuit breaker is mainly used for the display of the opening and closing status of the circuit breaker, which is connected in the control circuit of the circuit breaker to control or interlock its related electrical appliances through the opening and closing of the circuit breaker, such as outputting signal to signal lamp, relay, etc. Universal circuit breaker has six pairs of contacts (three normally open and three normally closed), and DW45 has eight pairs of contacts (four normally open and four normally closed). The rated current of the molded case circuit breaker is 100A, which is a single break point conversion contact. 225A and above are bridge contact structures, and the agreed heating current is 3A; The rated current of 400A and above can be installed with two normally open and two normally closed, and the agreed heating current is 6A. The number of operation performance is the same as the total number of operation performance of the circuit breaker.

It is used for the alarm contact of circuit breaker accident, and this contact can only operate when the circuit breaker trips. It is mainly used for the free tripping when the load of the circuit breaker has overload short circuit or undervoltage fault. The alarm contact is changed from the original normally open position to the closed position, and the indicator light, bell, buzzer, etc. in the auxiliary circuit are connected to display or remind the fault trip status of the circuit breaker. Since the probability of free tripping of the circuit breaker due to load fault is not too much, the service life of the alarm contact is 1/10 of the service life of the circuit breaker. Generally, the working current of alarm contact shall not exceed 1A.

Release

The tripper is a kind of tripper which is excited by voltage source. Its voltage is independent of the main circuit voltage. Shunt release is a kind of accessory for remote control. When the power supply voltage is equal to any voltage between 70% and 110% of the rated control power supply voltage, the circuit breaker can be reliably disconnected. The shunt release is a short-time working system, and the coil power on time generally cannot exceed 1s, otherwise the line will be burnt. In order to prevent the coil from burning out, a microswitch is connected in series with the shunt release coil. When the shunt release is closed by armature, the microswitch changes from normally closed state to normally open state. Because the control circuit of the shunt release power supply is cut off, even if the button is pressed and held artificially, the shunt release coil will not be energized all the time to avoid the coil burning. When the circuit breaker is reclosed, the microswitch will be in the normally closed position again. However, the universal DW45 product should be connected with a set of normally open contacts in series before the shunt release coil when it is used.

Undervoltage release is a kind of release which makes the circuit breaker open with or without delay when its terminal voltage drops to a specified range. When the power supply voltage drops (or even slowly drops) to 35%~70% of the rated working voltage, the undervoltage release shall operate. When the power supply voltage is equal to 35% of the rated working voltage of the release, the undervoltage release shall be able to prevent the circuit breaker from closing completely; When the power supply voltage is equal to or greater than 85% of the rated working voltage of the under voltage release, it shall be able to ensure the reliable closing of the circuit breaker under the hot condition. Therefore, when a certain voltage drop occurs in the power supply of the protected circuit, the circuit breaker can be automatically disconnected to cut off the power supply, so that the load electrical appliances or electrical equipment below the circuit breaker can be protected from under voltage damage. When in use, the under voltage release coil is connected to the power side of the circuit breaker, and the circuit breaker can be closed only after the under voltage release is powered on, otherwise the circuit breaker cannot be closed.

Practices

(1) Observe the appearance of LV circuit breaker, and check the label and technical parameters.

(2) Think about how to connect wires, and write the corresponding wire number and function of terminals in Table 1-1.

Table 1-1 Terminal function of LV circuit breaker

No.	Terminal name	Function	No.	Terminal name	Function
1			5		
2			6		
3			7		
4			8		

1.1.3.2 Fuse

Fuse refers to a kind of electrical appliance which breaks the circuit by fusing the melt with the heat generated by itself when the current exceeds the specified value. Fuse is a kind of current protector which is made according to the principle that after the current exceeds the specified value for a period of time, and the fuse melts the melt with the heat generated by itself, thus breaking the circuit. Fuse is widely used in high and low voltage distribution system, control system and electrical equipment. As the protector of short circuit and over-current, fuse is one of the most commonly used protective devices. The actual drawing and electrical symbols are shown in Figures 1-3 and 1-4 respectively.

Figure 1-3 Fuse physical

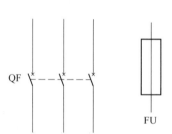

Figure 1-4 Fuse electrical symbol

Structural Characteristics

The rated current of the melt is not equal to the rated current of the fuse. The rated current of the melt shall be selected according to the load current of the protected equipment. The rated current of the fuse shall be greater than the rated current of the melt, which shall be determined in coordination with the main electrical appliance.

The fuse is mainly composed of melt, shell and support, in which melt is the key element to control the fusing characteristics. The material, size and shape of the melt determine the fusing characteristics. There are two kinds of melts: low melting point and high melting point. Low melting point materials (such as lead and lead alloy), have low melting point and they are easy to fuse. Due to their high resistivity, the cross-section size of the melt is large, and there are many metal vapors produced during the fusing, so they are only suitable for fuses with low breaking capacity. High melting point materials (such as copper and silver), have high melting point and are not easy to fuse. However, due to their low resistivity, they can be made into a smaller cross-section size than low melting point melts, and produce less metal vapor when they are fused, so they are suitable for fuses with high breaking capacity. The shape of the melt can be divided into two types: filament and ribbon. Changing the shape of the section can significantly change the fuse characteristics. Fuse has different fuse characteristic curves, which can be applied to different types of protection objects. It is used for the alarm contact of circuit breaker accident, and this contact can only operate when the circuit breaker trips. It is mainly used for the free tripping when the load of the circuit breaker has overload short circuit or undervoltage fault. The alarm contact is changed from the original normally open position to the closed position, and the indicator light, bell, buzzer, etc. in the auxiliary circuit are connected to display or remind the fault trip status of the circuit breaker. Since the probability of free tripping of the circuit breaker, due to load fault, is not too much, the service life of the alarm contact is 1/10 of the service life of the circuit breaker. Generally, the working current of alarm contact shall not exceed 1A.

The action of fuse is realized by fusing the melt. The fuse has a very obvious characteristic, which is ampere second characteristic. For the melt, its action current and action time characteristics are the ampere second characteristics of the fuse, also known as the inverse time delay characteristics. When the overload current is small, the fusing time is long; When the overload current is large, the fusing time is short.

Precautions for Use

The precautions of fuse for use are as follows:

(1) The protection characteristics of the fuse shall be suitable for the overload characteristics of the protected object. In consideration of the possible short circuit current, the fuse with corresponding breaking capacity shall be selected.

(2) The rated voltage of the fuse shall adapt to the line voltage level, and the rated current of the fuse shall be greater than or equal to the rated current of the melt.

(3) The rated current of fuse melts at all levels in the line shall be matched accordingly, and the rated current of the former level must be greater than that of the next level.

(4) The fuse melt shall use the matched melt according to the requirements, and it is not allowed to increase the melt at will or use other conductors to replace the melt.

Practice

Observe the appearance of the fuse, and check the label and technical parameters.

1.1.3.3 AC Contactor

Contactors are divided into AC contactors and DC contactors, which are used for power, distribution and electric field connection. Contactor is a kind of electric appliance that uses coil to flow through current to produce magnetic field and make contact closed to control load. The actual drawing and electrical symbols are shown in Figures 1-5 and 1-6 respectively.

Figure 1-5 Physical drawing of AC contactor

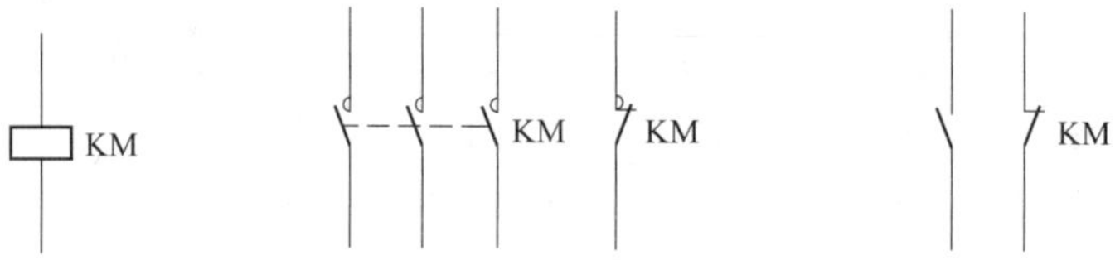

Figure 1-6 Electrical symbols of AC contactor

The working principle of the contactor is that, when the contactor coil is powered on, the coil current will generate a magnetic field, which makes the static iron core generate electromagnetic attraction to attract the moving iron core, and drives the AC contactor point to act. The normally closed contact is open, the normally open contact is closed, and the two are linked. When the coil is de energized, the electromagnetic attraction disappears, and the armature is released under the action of the release spring, so that the contact is restored. The normally open contact is opened, and the normally closed contact is closed. The working principle of DC contactor is similar to that of temperature switch.

The AC contactor uses the main contact to control the circuit and the auxiliary contact to conduct the control circuit. The main contact is normally open, while the auxiliary contact usually has two pairs of normally open and normally closed contacts. Small contactors are often used as intermediate relays to cooperate with the main circuit. The contact of AC contactor is made of silver tungsten alloy, which has good conductivity and high temperature ablation resistance. The power of AC contactor action comes from the magnetic field generated by AC through the coil with iron core. The electromagnet core is composed of two 'mountain' shaped (in Chinese) small silicon steel sheets, one of which is fixed iron core with a coil, and the working voltage can be selected in many ways. In order to stabilize the magnetic force, a short-circuit ring is added to the closing surface of the iron core. After the AC contactor loses power, it relies on the spring to return. The other half is the movable iron core, which has the same structure as the fixed iron core to drive the closing and opening of the main contact and the auxiliary contact. The contactors above 20A are equipped with arc extinguishing cover, which can quickly break the arc and protect the contact by using the electromagnetic force generated when the circuit is disconnected.

Practices

(1) Observe the appearance of AC contactor, and check the label and technical parameters.

(2) Think about how to wire, and write the wire number and function corresponding to the terminal in Table 1-2.

Table 1-2 **Function table of AC contactor terminal**

No.	Terminal name	Function	No.	Terminal name	Function
1			4		
2			5		
3			6		

Continued Table 1-2

No.	Terminal name	Function	No.	Terminal name	Function
7			11		
8			12		
9			13		
10			14		

1.1.3.4 Button

Button is a kind of commonly used control electrical element, which is used to connect or disconnect the control circuit (in which the current is very small), so as to achieve the purpose of controlling the operation of motor or other electrical equipment. Buttons are divided into:

(1) Normally open button: button with switch contacts disconnected.

(2) Normally closed button: button with switch contacts connected.

(3) Normally open and normally closed button: button with switch contacts both connected and disconnected.

(4) Action click button: mouse click button, also known as a key. It is a kind of switch (or switch), used to control some mechanical or program functions.

Generally speaking, the red button is used to stop the function, while the green button can start a function. The shape of the button is usually round or square. Most electronic products are useful to the most basic human-computer interface tool, button. With the improvement and innovation of industrial level, the appearance of button is becoming more and more diversified and rich in visual effects. The actual drawing and electrical symbols are shown in Figures 1-7 and 1-8 respectively.

Figure 1-7　Physical figure of button

Figure 1-8　Electrical symbol of button

Practices

(1) Observe the appearance of the button, and check the label and technical parameters.

(2) Think about how to connect wires, and write the corresponding wire number and function of terminals in Table 1-3.

Table 1-3 Function table of button terminal

No.	Terminal name	Function	No.	Terminal name	Function
1			4		
2			5		
3			6		

1.1.3.5 Three Phase Asynchronous Motor

Three phase asynchronous motor is a kind of induction motor, which is powered by 380V three-phase alternating current (phase difference of 120°). Because the rotor and stator rotating magnetic field of three-phase asynchronous motor rotate in the same direction and at different speeds, there is a slip rate, so it is called three-phase asynchronous motor. The rotor speed of three-phase asynchronous motor is lower than that of rotating magnetic field. Because of the relative motion between rotor winding and magnetic field, electromotive force and current are generated, and electromagnetic torque is generated by the interaction between rotor winding and magnetic field to realize energy transformation.

Compared with single-phase asynchronous motor, three-phase asynchronous motor has better performance and can save various materials. According to different rotor structure, three-phase asynchronous motor can be divided into cage type and winding type. Cage rotor asynchronous motor has been widely used because of its simple structure, reliable operation, light weight and low price. Its main disadvantage is the difficulty of speed regulation. The rotor and stator of wound three-phase asynchronous motor are also provided with three-phase winding and connected with external rheostat through slip ring and brush. Adjusting the rheostat resistance can improve the starting performance and speed of the motor.

Structure of Three-phase Asynchronous Motor

Three-phase asynchronous motor is a kind of rotating motor which transforms alternating current energy into mechanical energy based on electromagnetic principle. The basic structure of three-phase squirrel cage asynchronous motor consists of stator and rotor. The stator, which is the static part of the motor, is mainly composed of stator core, three-phase symmetrical stator winding and frame. Generally, the three-phase stator winding has six outgoing lines, and the outgoing end is installed in the junction box outside the machine base, as shown in Figure 1-9. According to the different voltage of the three-phase power supply, the three-phase stator winding can be connected into a star (Y) or a triangle (△), as shown in Figure 1-10, and then connected with the three-phase AC power supply. The rotor is mainly composed of rotor core, rotating shaft, squirrel cage rotor winding, fan, etc. It is the rotating part of the motor. The rotor winding of small capacity squirrel cage asynchronous motor is large. They are all cast with aluminum, and the cooling mode is generally fan cooling.

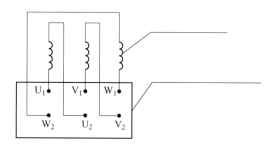

Figure 1-9 Three-phase stator winding diagram

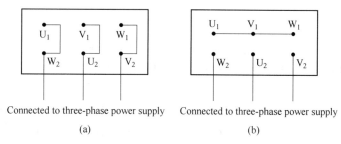

Figure 1-10 Three-phase stator winding connection
(a) Delta connection; (b) Y connection

Nameplate of Three-phase Asynchronous Motor

The rating of three-phase asynchronous motor is marked on the motor nameplate, and the nameplate of three-phase squirrel cage asynchronous motor is shown in Table 1-4.

Table 1-4 Nameplate of three-phase asynchronous motor

Model	PN/W	UN/V	IN/A
YS6324	180	380/660	0.65/0.38
NN/rpm	Connection method	No.	Insulation class
1400	Y/△	—	B

Note: 1. Power: The mechanical power output on the motor shaft under rated operation.
  2. Voltage: Under rated operation, the line voltage value of power supply shall be added to the stator three-phase winding.
  3. Connection method: When the rated voltage is 380V/220V, it shall be Y/△ connection method.
  4. Current: Under rated operation, when the motor output rated power, the line current of stator circuit.

Inspection of Three-phase Asynchronous Motor

Necessary inspection shall be made before using the motor, as follows:

(1) Mechanical inspection. It should check whether the outgoing line is complete and reliable, whether the rotor rotates flexibly and evenly, and whether there is abnormal sound, etc.

(2) Electrical inspection. In the daily operation of the motor, the coil is often loose, which makes the insulation wear and aging, or the surface is polluted, damped and so on, which causes the insulation resistance to decline day by day. The reduction of the insulation resistance to a certain value will affect the starting and normal operation of the motor, and even damage the motor, endangering personal safety. Therefore, the insulation resistance of each phase winding to the casing and the insulation resistance between windings shall be measured before the use of various motors or after the mould season, damp and reinstallation. The measurement of insulation resistance is generally carried out with a megger. Learning how to use the megger can bring us convenience when checking the insulation of motor, electrical appliance and circuit and measuring high value resistance.

A megger is used to check the insulation between the windings of the motor and between the windings and the housing. And the megger is also used to test the insulation resistance value of U_1, V_1 and W_1 at the beginning of each phase winding to the casings and between each phase winding. During the measurement, the grounding terminal of the megger is connected to the housing (pay attention not to touch the paint, so as to avoid inaccurate measurement data). The other test terminal is connected to U_1, V_1 and W_1 ends of the stator winding respectively. Then the handle of the megger is rotated at a certain speed (generally 120rpm), the handle speed is kept unchanged, and the reading of the megger is read. If the value is more than $0.5M\Omega$, it is qualified; If it is less than $0.5M\Omega$, it means that the insulation resistance of the motor decreases, and the motor can not be used again without maintenance or drying treatment. To test the insulation resistance between the two-phase windings, the two test ends of the megger are just connected to the beginning of any two-phase winding, the handle of the megger is rotated in the same way as above, and the reading of the megger is read out. When the megger is shaken, the voltage between its two test terminals can reach 500V, so pay attention to safety during the test. The wiring method is shown in Figure 1-11.

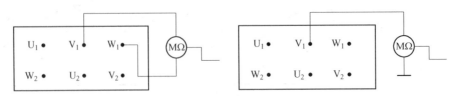

Figure 1-11 Using megger to check the insulation performance between motor windings and between windings and casings

Practices

(1) Observe the appearance of three-phase asynchronous motor, and check the label and technical parameters.

(2) Think about how to wire, and write the corresponding wire number and function of the terminal in Table 1-5.

Table 1-5 Three-phase asynchronous motor terminal function table

No.	Terminal name	Function	No.	Terminal name	Function
1			5		
2			6		
3			7		
4			8		

1.1.4 Task Implementation

Observe the schematic shown in Figure 1-12, and then:
(1) Describe its function.
(2) Draw schematic diagram.
(3) Mark line number.

In the process of implementation, if there is a fault, it is necessary to find, analyze and eliminate the fault. Pay attention to the cultivation of professional core literacy such as safety awareness and team awareness. Finally, analyze and summarize the whole task and fill in Table 1-7.

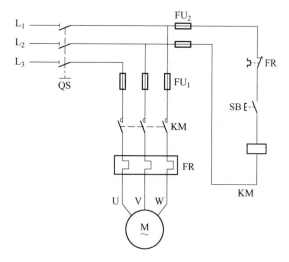

Figure 1-12 Inching circuit diagram of relay system control

1.1.5 Task Evaluation

The scoring table is shown in Table 1-6.

Table 1-6 Scoring table

Task content	Assessment requirements	Scoring criteria	Allotment	Points deducted	Score
Preparation	Prepare well before the task is implemented	Preparatory work is the entry stage of the whole implementation process. Only after this stage is completed, can the next step be carried out. If this step is not done or fails, the whole project will be scored 0 point	10		
Description of working principle	Correctly explain the working principle of the whole system	(1) 10 points will be plused for the description of main circuit function; (2) 10 points will be plused for the description of function description of control	20		
Schematic drawing	Complete the schematic drawing correctly	(1) 5 points will be deducted for each line with drawing error; (2) 10 points will be plused for the principle drawing neat; (3) 10 points will be plused for the schematic scale appropriate	30		
Wire number tagging	Complete the marking of schematic line number correctly	(1) 5 points will be deducted for each error of wire number marking; (2) 5 points will be deducted for each missing mark of line number; (3) 5 points will be deducted for each place where the wire number does not match the actual component	30		
Safety and civilization production	Comply with 8s management system and safety management system	(1) 10 points will be deducted if no labor protection articles are worn; (2) If there is potential safety hazard in operation, 5 points will be deducted each time until all points are deducted; (3) If the operation site is not cleaned up in time, 2 points shall be deducted for each time until all points are deducted	10		
		Total score			

1.1.6 Task Summary

The task analysis summary record is shown in Table 1-7.

Table 1-7 Task analysis summary record

		Task analysis summary record
Fault 1	Fault Phenomenon	
	Cause of failure	
	Exclusion process	
Fault 2	Fault phenomenon	
	Cause of failure	
	Exclusion process	
Fault 3	Fault phenomenon	
	Cause of failure	
	Exclusion process	
	Summary	

1.1.7 Task Development

Task requirements: According to Figure 1-12, complete the line connection and analyze the whole process.

Task 1.2 PLC Basics

1.2.1 Task Description

Programmable logic controller (PLC) is a kind of digital operation electronic system specially designed for application in industrial environment. It uses a kind of programmable memory, in which the instructions of logic operation, sequence control, timing, counting and arithmetic operation are stored, and various types of mechanical equipment or production process are controlled by digital or analog input. Through the study of this task, PLC will be preliminary understood, including generation, definition, feature development, etc., laying a good foundation for subsequent PLC programming.

1.2.2 Task Target

(1) Understand the generation and definition of PLC.
(2) Understand the characteristics of PLC.
(3) Understand the development direction of PLC.
(4) Learn about the 1200 series PLC.
(5) Master the wiring mode of PLC input terminal.
(6) Train students' sense of safe operation and team work.

1.2.3 Task-related Knowledge

This course uses Siemens 1215C series PLC. Programmable controller is a kind of general industrial control computer. It is based on microprocessor and developed by using computer technology, microelectronics technology, automatic control technology, digital technology and network communication technology. It has become one of the four pillars of modern industrial control (PLC technology, robot technology, CAD/CAM, and numerical control technology), which are process oriented, user-oriented, industrial environment-friendly, easy to operate and high reliability. Its control technology represents the advanced level of current program control, and it has become the basic device of automatic control system.

The original programmable logic controller is mainly logic control, so it is called PLC for short. Now the function of PLC is expanding constantly. In addition to logic control, it also adds functions such as analog quantity adjustment, numerical calculation, monitoring, communication networking, etc. So it is renamed as PLC (PC for short). But in order to be different from personal computer, many people call it PLC for short.

1.2.3.1 Generation and Definition of PLC

At the end of 1960s, most of industrial production was based on large-scale production with few varieties, and the control of this large-scale production line was dominated by relay control system. Due to the development of the market, the development direction of industrial production is required to change the production mode of small batch and multiple varieties. In this way, the relay control system needs to be redesigned and installed, which is very time-consuming, labor-intensive and material-intensive, hindering the shortening of the renewal cycle. In order to change this situation, in 1968, General Motors (GM) of the United States opened a public tender, expecting to design a new type of automatic industrial control device to replace the relay control device, so as to achieve the purpose of constantly updating the vehicle model. The following 10 indicators are proposed:

(1) The program is convenient and can be modified on site.
(2) It is easy to maintain and adopts plug-in structure.
(3) Reliability is higher than relay control device.
(4) The data can be sent directly to the management computer.
(5) The input can be 115V AC.
(6) The output is AC 115V, 2A or above, which can drive solenoid valve and contactor directly.
(7) User storage capacity can be expanded to at least 4kB.
(8) The volume is smaller than the relay control device.
(9) When expanding, the original system changes little.
(10) Cost can compete with relay control.

In 1969, DEC developed the world first programmable logic controller (PLC) according to the requirements of the bidding, and it was applied to the automatic assembly line of General Motors Company of America, and it was successful. Since then, PLC has been widely used in other industrial fields in the United States, creating a new era of industrial control.

The International Electrotechnical Commission (IEC) has issued the standard draft of PLC for three times successively. The definition of PLC in the third draft in February 1987 is: 'PLC is an electronic system of digital operation, designed for application in industrial environment. It uses programmable memory to store and execute logical operation and sequence in its interior-control, timing, counting and arithmetic operation instructions, and control the production process of various types of machinery through digital and analog input and output. The programmable controller and its related peripheral equipment shall be designed according to the principle of being easy to connect with the industrial system as a whole and expand its functions.' It can be seen from the above definition that PLC is a computer at first, and it is an industrial computer specially designed for application in industrial environment. Because it is designed and produced according to the index proposed by the user, and at the same time, it has unique characteristics that other industrial control equipment is difficult to have, making it widely adapt to various industrial controls.

1.2.3.2 Characteristics of PLC

High Reliability and Strong Anti-interference Ability

PLC is specially designed for industrial control environment, so high reliability and strong anti-interference are one of its main characteristics. As a result of a series of measures taken in the design, the average trouble free working time of PLC is several hundred thousand hours. It can be said that up to now, the reliability of any industrial control equipment can not reach the level of PLC, and with the improvement of device technology, the reliability of PLC will continue to improve.

Generally, the causes of digital electronic equipment failure can be divided into two categories: One is caused by bad external environment, such as electromagnetic interference, high temperature, high humidity, vibration, and harmful gas; The other is caused by aging and failure of internal devices, loss of memory information, operation of wrong programs, etc. Therefore, measures can be taken from both hardware and software to reduce the occurrence of faults.

The hardware of PLC adopts modular structure, and take anti-interference measures such as photoelectric isolation, shielding, filtering, etc. In addition, it can set output interlock protection, environmental detection, self diagnosis circuit, etc. for some modules.

The software of PLC sets self diagnosis, real-time monitoring, information protection and other programs. In addition, dual CPU redundant system or CPU voting system is adopted in large PLC to further improve the reliability of PLC.

Easy to Program and Use

This is another major feature of PLC. PLC mostly uses ladder diagram similar to relay control circuit for programming, which is intuitive, easy to master by general engineering technicians, and easy to operate and use.

PLC also designed other kinds of programming languages, such as instruction statement table programming language, function block programming language and so on, to meet the needs of different programmers and better complete various control functions.

Strong Function, Good Universality and Flexible Use

Modern PLC uses the latest technologies such as computer technology, microelectronics technology, digital technology, network communication technology and integrated technology to enhance its complex control and communication networking functions. Moreover, the current PLC products have realized serialization, modularization and standardization. It can also flexibly and conveniently form control systems of different scales and functions to meet the needs of users.

Easy to Install and Debug

As PLC has complete functions, as long as it can reasonably select various modules to make up the system, there is no need for additional hardware configuration, and no need for secondary development of software. The application program of PLC can also be easily simulated and debugged in the laboratory. After successful debugging, it can be installed and debugged on site.

Small Size, Light Weight and Low Power Consumption

Because PLC adopts microelectronic technology, it is small in size, compact in structure, light in weight and low in power consumption.

1.2.3.3 Direction Development of PLC

With the development of computer technology, digital technology, semiconductor integration technology, network communication technology and other high-tech, PLC has also been rapid development. At present, PLC has been widely used in various fields.

The development direction of PLC can be roughly divided into two parts: One is to develop to the miniaturization direction with smaller volume, stronger function and lower price to provide the miniaturization PLC control system with higher performance price ratio, so as to make its application scope more extensive; The other is to develop to the large-scale direction with faster speed, more functions, stronger networking and communication ability to provide the high performance, high speed and high cost performance medium PLC control system to meet the needs of large-scale and complex control system, specifically reflected in the following aspects:

(1) Enhanced network communication.

(2) Develop intelligent I/O modules.

(3) Adopt various programming languages.

(4) Enhance the ability of external fault detection and handling.

1.2.3.4 Introduction of 1200 Series PLC

SIMATIC S7-1200 is a compact and modular PLC, which can complete simple logic control, advanced logic control, HMI and network communication tasks. It is easy to design and implement the automation system which needs network communication function and single screen or multi screen HMI. It has the advanced application functions of supporting small motion control system and process control system. The new modular SIMATIC S7-1200 controller is the core of Siemens, which can realize simple but highly accurate automation tasks. SIMATIC S7 - 1200 controller realizes modular and compact design, with powerful function, safe investment and fully suitable for various applications. The design with strong expansibility and flexibility can realize the communication interface of the highest standard industrial communication and a set of powerful integrated technical functions, making an important part of a complete and comprehensive automation solution. Its appearance is shown in Figure 1-13.

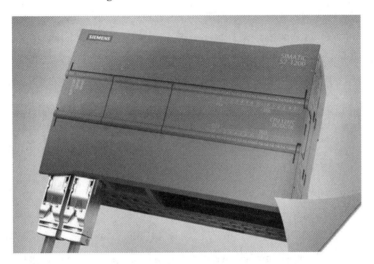

Figure 1-13 Appearance of Siemens 1200 series PLC

Advantages

The performance of the SIMATIC HMI foundation panel is optimized to be perfectly compatible with this new controller and powerful integrated engineering configuration, ensuring simplified development, fast start-up, precise monitoring and highest level of availability. It is the synergy and innovative functions between these products that help you to improve the efficiency of small automation systems to an unprecedented level.

It realizes a simple and effective technical task solution, and fully meets the application requirements of a series of independent automation systems. In order to achieve the highest efficiency in engineering configuration, SIMATIC S7-1200 is programmed with SIMATIC WinCC basic and a new fully integrated engineering configuration SIMATIC STEP 7 basic. The design concept of SIMATIC STEP 7 basic is intuitive, and easy to learn and use. This design concept enables you to achieve maximum efficiency in engineering configurations. Some intelligent functions, such as visual editor, drag and drop function and 'IntelliSense' tool, can make your project go on more quickly. The architecture of this new software stems from the constant pursuit of future innovation. Siemens has many years of experience in the field of software development, so the design of SIMATIC STEP 7 is future oriented.

Design and Function

SIMATIC S7-1200 system has five different modules: CPU 1211C, CPU 1212C, CPU 1214C, CPU 1215C and CPU 1217C. Each of these modules can be expanded to fully meet your system needs. A signal board can be added in front of any CPU to easily expand digital or analog I/O without affecting the actual size of the controller. The signal module can be connected to the right side of the CPU to further expand the digital or analog I/O capacity. CPU 1212C can be connected with 2 signal modules, and CPU 1214C, CPU 1215C and CPU 1217C can be connected with 8 signal modules. Finally, the left side of all SIMATIC S7-1200 CPU controllers can be connected with up to 3 communication modules to facilitate end-to-end serial communication.

The utility model has the advantages of simple and convenient installation, space saving design, etc. All SIMATIC S7-1200 hardware has built-in snaps that can be easily installed on standard 35mm DIN rails. These built-in snaps can also be clipped into extended positions, and mounting holes can be provided when the panel needs to be installed. SIMATIC S7-1200 hardware can be installed in horizontal or vertical position, providing you with other installation options. These integrated functions provide maximum flexibility for users in the installation process, and make SIMATIC S7-1200 provide practical solutions for various applications.

All SIMATIC S7-1200 hardware is specially designed to save control panel space. For example, after measurement, the width of CPU 1214C is only 110mm, and the width of CPU 1212C and CPU 1211C are only 90mm. Combined with the small space occupied by the communication module and the signal module, the modular compact system saves precious space during the installation process, providing you with the highest efficiency and maximum flexibility. SIMATIC S7-1200 is a scalable, compact and automatic modular concept, which can realize simple communication and effective technical task solutions, and fully meet a series of independent automation requirements.

1.2.3.5 Power Supply and Input Terminal Wiring of PLC

The external wiring diagram of CPU 1215C DC/DC/DC (6ES7 215-1AG40-0XB0) is shown in Figure 1-14. It can be seen from the figure that the power supply of PLC is DC 24V, in which L+ is connected to the positive pole of power supply and M is connected to the negative pole of power supply. The common terminal of 1M input forms a loop after connecting with input terminal, input switch and DC 24V power supply; 2M is the common terminal of analog output; 3M is the common terminal of analog input; 4M is the common terminal of PLC output.

Practices

(1) Observe PLC on XK-SX2C advanced maintenance electrician training platform, and check label and technical parameters.

(2) Think about how to connect wires, and write the corresponding wire number and function of terminals in Table 1-8.

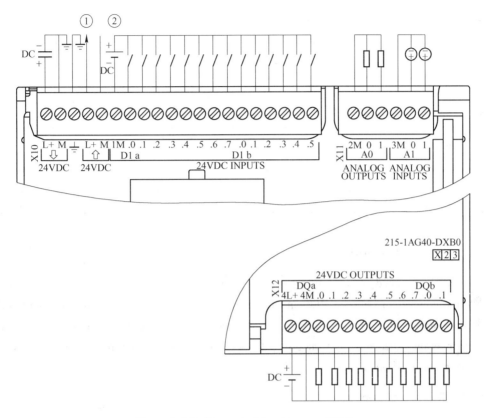

Figure 1-14 External wiring diagram of PLC

Table 1-8 Terminal function table of PLC

No.	Terminal name	Function	No.	Terminal name	Function
1			15		
2			16		
3			17		
4			18		
5			19		
6			20		
7			21		
8			22		
9			23		
10			24		
11			25		
12			26		
13			27		
14			28		

Continued Table 1-8

No.	Terminal name	Function	No.	Terminal name	Function
29			37		
30			38		
31			39		
32			40		
33			41		
34			42		
35			43		
36			44		

1.2.4 Task Implementation

On XK-SX2C advanced maintenance electrician training platform, the jumper wire is used to complete the connection between PLC and power supply, and the power supply is connected on the basis of correct connection, and then two switches are connected. On the basis of ensuring the correct connection, the power is turned on, the switch is pressed, and the operation of the indicator light is observed on the PLC.

1.2.4.1 Task Requirements

Refer to Figure 1-14 for power on test of PLC, connect two buttons for test, and then score the project after completion, as shown as Table 1-9.

1.2.4.2 Operation Steps

The operation steps are as follows:
(1) Find out the fault of the required components to ensure the normal use of each component.
(2) Draw the schematic diagram and mark the line number.
(3) Connect the power line of PLC and test with Multimeter.
(4) Connect two switch buttons.
(5) Check with multimeter before power on.
(6) Power on to observe the operation status of the system.
(7) Summarize and record.

1.2.4.3 Analysis and Summary

Summarize the whole task implementation process, find out the deficiencies, especially pay attention to the faults, and record in Table 1-10 after analysis and summary.

1.2.5 Task Evaluation

The scoring table are shown in Table 1-9.

Table 1-9 Scoring Table

Task content	Assessment requirements	Scoring criteria	Allotment	Points deducted	Score
Preparation	(1) Find out the fault of the required components to ensure the normal use of each component; (2) Draw schematic diagram	Preparatory work is the entry stage of the whole implementation process. Only after this stage is completed can the next step be carried out. If this step is not done or fails, the whole project will be scored 0 point	10		
Connection of power cord	Be able to supply power to PLC correctly	(1) 5 points will be deducted if the positive connection is not correct; (2) 5 points will be deducted if the negative connection is incorrect	10		
Connection of input switch button	Connect the two input switch buttons correctly	(1) 10 points will be deducted for each wrong connection; (2) 10 points will be deducted if loop cannot be formed; (3) 5 points will be deducted for each connection error of normally open and normally closed contacts.	30		
Safety test before power on	Before powering up PLC, safety test shall be carried out to ensure no problem before powering up	Make sure there is no short circuit, open circuit, virtual connection and other phenomena. 0 point will not be given for the whole project in this step	10		
Power on operation demonstration	Correctly demonstrate the input switch button, and can be explained with hardware	5 points will be deducted for each place where the operation phenomenon cannot be analyzed and explained	30		
Safety and civilization production	Comply with 8S management system and safety management system	(1) 10 points will be deducted if no labor protection articles are worn; (2) If there is potential safety hazard in operation, 5 points will be deducted each time until all points are deducted; (3) If the operation site is not cleaned up in time, 2 points will be deducted for each time until all points are deducted	10		
		Total score			

1.2.6 Task Summary

The task analysis summary record is shown in Table 1-10.

Table 1-10 Task analysis summary record

		Task analysis summary record
Fault 1	Fault phenomenon	
	Cause of failure	
	Exclusion process	
Fault 2	Fault phenomenon	
	Cause of failure	
	Exclusion process	
Fault 3	Fault phenomenon	
	Cause of failure	
	Exclusion process	
	Summary	

1.2.7 Task Development

Task requirements: Learn 8s management system and use it in actual study and work. 8s is the eight items of seiri, seiton, seiso, setketsu, shitsuke, safety, save and study. Because the ancient Roman pronunciation starts with 'S', it is called 8s for short. The purpose of 8s management method is to improve the quality of corporate culture, eliminate safety hazards, and save cost and time by establishing learning organization on the basis of on-site management. So that enterprises in the fierce competition are always in an invincible position.

1.2.7.1 Seiri

Seiri refers to cleaning up the chaotic state into an orderly state. The purposes of Seiri are as follows:

(1) Make space and increase the working area.

(2) Ensure smooth logistics, and prevent misuse and misdelivery.

(3) Create a fresh workplace.

The key points of implementation are as follows:

(1) Conduct a comprehensive inspection of your workplace (SCOPE), including visible and invisible ones.

(2) Establish a judgment standard of 'Want' and 'Don't'.

(3) Be determined to remove the unwanted items from the workplace.

(4) Investigate the use frequency of the required items and determine the daily consumption and placement location.

(5) Formulate waste treatment methods.

(6) Conduct daily self inspection.

1.2.7.2 Seiton

Seiton refers to scientifically and reasonably arranging and placing the items left on the production site, after finishing the previous step, so as to obtain the required items with the fastest speed and complete the operation under the most effective rules, regulations and the most simple process.

The purposes of Seiton as follows:

(1) Make the workplace clear at a glance and create a neat working environment.

(2) Without wasting time to find things, you can find things in 30s and use them immediately.

The key points of implementation are as follows:

(1) The work of the previous step should be carried out.

(2) The process layout should be made sure of the location and quantity of the items as follows:

1) The location of the items should be set 100% in principle;

2) the storage of the items should be fixed point (where is appropriate), fixed volume (what container and color is used), and fixed quantity (specify the appropriate quantity);

3) Only the items that are really needed should be placed near the production line.

(3) Specified placing methods, as follows:

1) Easy to take and improve efficiency;

2) Within the specified range;

3) More work on the placing method.

(4) Marking and positioning.

(5) Place and item.

The identifications of rectification are as follows:

(1) Place and item identification shall correspond one by one in principle.

(2) Identification method shall be unified throughout the company.

1.2.7.3 Seiso

Seiso refers to removing the dirt in the workplace, prevent the occurrence of pollution, and keeping the post in a clean and tidy state without garbage and dust. Objects of cleaning include floor, wall, workbench, tool rack, tool cabinet, etc., machines, tools, measuring tools, etc.

The purposes of Seiso are as follows:

(1) Eliminate dirt, keep the workplace clean and bright, so that employees can maintain a good working mood.

(2) Stabilize quality, and ultimately achieve zero failure and zero loss of production.

The key points of implementation are as follows:

(1) Establish cleaning responsibility area (inside and outside the work area).
(2) Carry out routine cleaning, clean up dirt, and form responsibility and system.
(3) Investigate pollution sources, and put an end to or isolate them.
(4) Establish cleaning benchmark as a standard.

1.2.7.4 Seiketsu

Seiketsu refers to carrying out the above 3S (sorting, rectifying and cleaning) implementation to the end, forming a system, and implementing and maintaining the results.

The purposes of Seiketsu are to maintain the above 3S results and show the 'Abnormal' place.

The key points of implementation are as follows:
(1) The first 3S work shall be thoroughly implemented.
(2) Regular inspection shall be carried out to implement the reward and punishment system and strengthen the implementation.
(3) Management personnel shall often take the lead in patrol inspection to show their attention.

1.2.7.5 Shitsuke

Literacy means that everyone should act according to the regulations, and form a good habit of 8S management at any time and stick to it.

The purposes of literacy are as follows:
(1) Improve the quality of employees and cultivate them to be a person who abides by the rules and regulations and has good work quality habits.
(2) Build a group spirit.

The key points of implementation are as follows:
(1) Relevant rules and regulations to be followed by the training.
(2) Strengthen education and practice for new employees.

1.2.7.6 Safety

Safety refers to removing potential safety hazards, ensure personal safety, product quality safety of employees on the job site, and prevent accidents.

The purposes of safety are as follows:
(1) Standardize operation, ensure product quality, and eliminate safety accidents.
(2) Ensure personal safety of employees, and ensure continuous and normal production.
(3) Reduce economic losses caused by safety accidents.

The key points of implementation are as follows:
(1) Formulate correct operation process, and timely supervise and guide.
(2) Timely find and eliminate the non safety factors, clean and repair all equipment, and find out the existing problems in advance, so as to eliminate the potential safety hazards.
(3) Thoroughly carry out safety activities at the operation site, so that the employees can

maintain the key points of safe power use, smooth passage and moving articles, make a habit, and establish a regular operation site.

(4) Employees use protective equipment correctly and do not work against rules.

1.2.7.7 Saving

Saving refers to the reasonable use of time, space, resources and other aspects to reduce waste and cost, so as to give full play to their maximum efficiency, and to create an efficient and best use workplace.

The purposes of saving are to cultivate the habit of reducing cost and the consciousness of reducing waste of operators.

The key points of implementation are as follows:

(1) Treat the resources of the enterprise with the attitude of being the owner.
(2) Make the best use of what can be used.
(3) Don't discard at will, and think about the remaining use value before discarding.
(4) Reduce the action waste and improve the operation efficiency.
(5) Strengthen the awareness of time management.

1.2.7.8 Studying

Studying refers to in-depth study of professional and technical knowledge. It can acquire knowledge from practice and books, and constantly learn from colleagues and supervisors.

The purposes of studying are as follows:

(1) Learn the advantages, and improve ourselves comprehensive quality.
(2) Enable employees to develop better, so as to generate new impetus for enterprises to cope with the possible future competition and change.

The key points for implementation are as follows:

(1) Learn all kinds of new skills and skills to continuously meet the needs of individual and company development.

(2) Share with others to achieve complementarity, mutual benefit, win-win manufacturing, weak complementary knowledge and technology, and defects of complementary ability, and improve the overall competitiveness and adaptability.

(3) Internal and external customer service awareness, for the interests of the collective (or individual) or work for your career and serve colleagues and customers related to you (for example, pay attention to the service of internal customers).

Task 1.3 Use of Programming Software

1.3.1 Task Description

TIA is the abbreviation of the fully integrated automation software TIA portal. The English full name is totally integrated automation portal. It is a new fully integrated automation software released by Siemens industrial automation group. It is the first automation software in the industry to adopt a unified engineering configuration and software project environment, which is applicable to almost all automation tasks. With the help of the new engineering software platform, users can develop and debug the automation system quickly and intuitively. As the basis of all future software engineering configuration packages, TIA can configure, program and debug all automation and drive products involved in Siemens fully integrated automation.

Through the learning of this task, master the ability to input, download and monitor the program in the TIA programming software. Summarize the problems encountered in the process of task implementation and record them in the corresponding tables.

1.3.2 Task Target

(1) Master the input method of software.

(2) Master the configuration method of hardware configuration equipment.

(3) Master the method of software monitoring.

(4) Ability to find, analyze and troubleshoot.

(5) Train students' sense of safe operation and team work.

1.3.3 Task-related Knowledge

The usage and steps of TIA programming software are as follows:

(1) Double click the botu software icon. In the botu software interface, click 'Create new project' on the left side of the map. You can change the name and path of the new project, and then click 'Create', as shown in Figure 1-15.

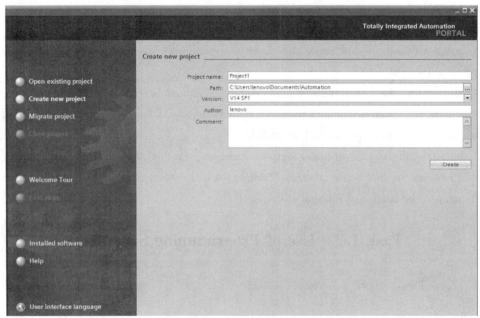

Figure 1-15 Creating a new project

(2) In the pop-up window, click 'Configure device' icon to configure the hardware device, as shown in Figure 1-16.

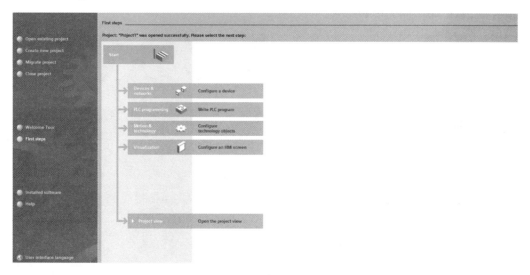

Figure 1-16 Hardware configuration

(3) In the window that appears, click 'Controller' icon. In the controller list that appears, select 'CPU 1215C DC/DC/DC', and select according to the actual version number of PLC, and then click the button 'Add', as shown in Figure 1-17.

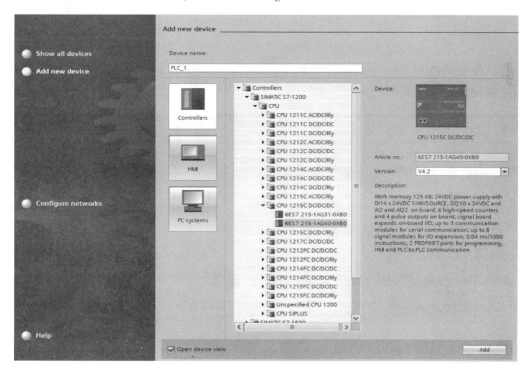

Figure 1-17 PLC hardware addition

(4) In the window that appears, click to select the Ethernet port of PLC, click 'Attribute' label in the lower part of the window, and click 'Ethernet address' on the left side to set the IP address of PLC, as shown in Figure 1-18.

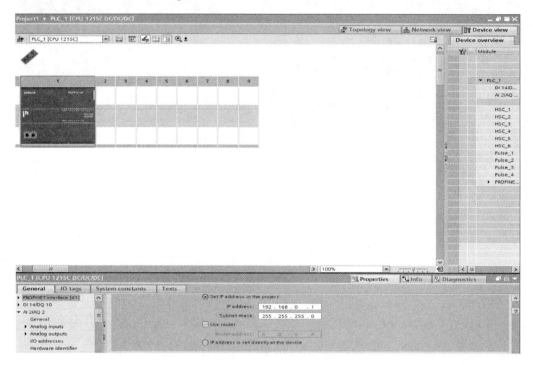

Figure 1-18   PLC address setting

(5) After selecting PLC, click the icon to download the hardware configuration to the PLC, as shown in Figure 1-19.

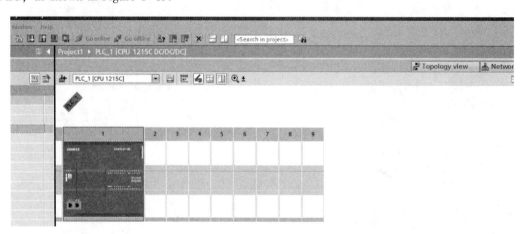

Figure 1-19   Hardware configuration download

(6) In the pop-up window, click 'Start search' button to find the target PLC and its IP ad-

dress, as shown in Figure 1-20.

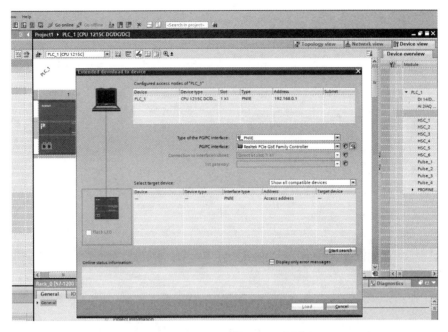

Figure 1-20   Finding target PLC

(7) After finding and connecting to the target PLC, click 'Load' button to download the hardware configuration to the PLC, as shown in Figure 1-21. During the download process, if the window appears as shown in Figure 1-22, click 'Continue under different circumstances' to continue the download.

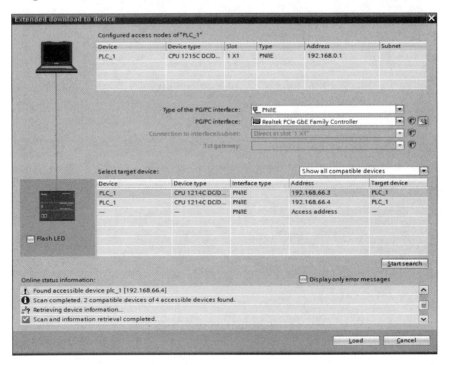

Figure 1-21   Connecting PLC and downloading (1)

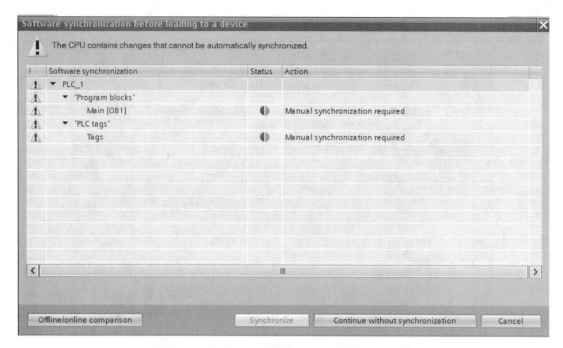

Figure 1-22　Connecting PLC and downloading (2)

(8) In the following window, select 'Stop module' option, and then click 'Load' to download the PLC hardware to the PLC, as shown in Figure 1-23.

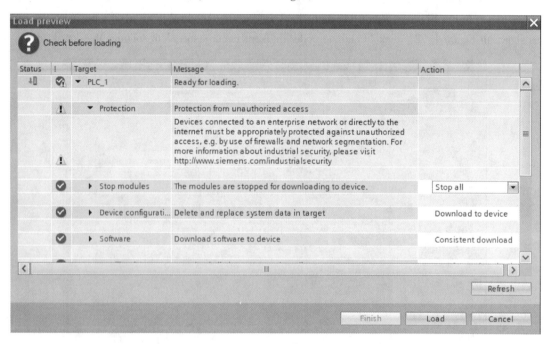

Figure 1-23　Pre download inspection

(9) After the hardware download, expand 'Program block' under the 'Equipment' tab, and then double click 'Main (OB1)' to open the programming interface, as shown in Figure 1-24.

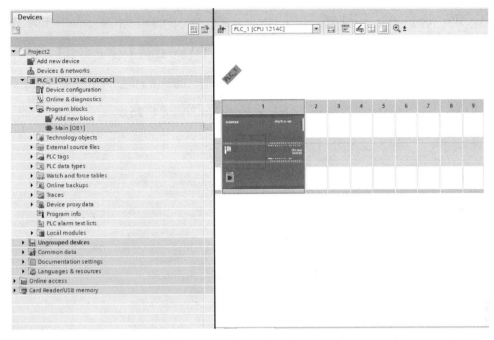

Figure 1-24 Open program editing interface

(10) In the open programming interface, you can drag or double click the corresponding instruction of area 1 or area 2, as shown in Figure 1-25, to add it to the logic line. For common instructions, you can drag the instruction of area 1, as shown in Figure 1-25, to area 2 to increase the convenience of programming. Logical lines can be added or deleted by icons [icon] or [icon].

Figure 1-25 Introduction of programming interface

(11) After the program is edited, click the icon ![icon] to download the program to PLC. At this time, if the PLC displays the stop state, you can click the icon ![icon] to start the PLC in the running state. For the convenience of debugging the program, you can click ![icon] to monitor the program and view the operation of the program, as shown in Figure 1-26.

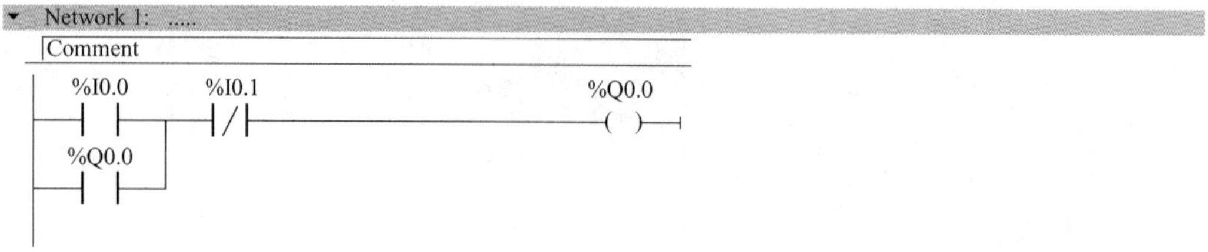

Figure 1-26 Program operation monitoring

Practice

Input the program in Figure 1-26, download it to PLC, and open the monitoring to observe the operation status.

1.3.4 Task Implementation

1.3.4.1 Task Requirements

The task requires good command of programming software. On XK-SX2C advanced maintenance electrician training platform, connect PLC according to the input point in Figure 1-27, and download the program. On the basis of correct connection, turn on the power, operate the input button switch, and observe the operation status of PLC. In the process of implementation, if there is a fault, it is necessary to find, analyze and eliminate the fault. Pay attention to the cultivation of professional core literacy such as safety awareness and team awareness.

Finally, score the project after completion, as shown in Table 1-12.

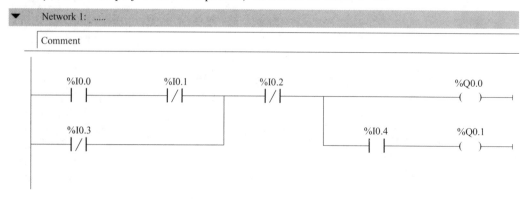

Figure 1-27 Reference procedure

1.3.4.2 Operation Steps

The operation steps are as follows:

(1) Find out the fault of the required components to ensure the normal use of each component.

(2) Draw schematic diagram.

(3) Connect the power line of PLC and measure it with multimeter to ensure no error before power on.

(4) Determine I/O address assignment and fill in Table 1-11.

Table 1-11 I/O address allocation

Input signal				Output signal			
No.	Function	Element	Address	No.	Control object	Element	Address
1				1			
2				2			
3				3			
4				4			
5				5			

(5) According to the I/O address assignment, complete the line connection of the input terminal.

(6) Power on PLC after checking.

(7) Download the program in Figure 1-27 to PLC for debugging.

(8) Use the multimeter to check before power on. In case of any fault, remove it in time and record the summary.

(9) Power on and observe the operation status of the system. In case of any fault, remove it in time and record the summary.

(10) Summarize and record.

1.3.4.3 Analysis and Summary

Summarize the whole task implementation process, find out the deficiencies, especially pay attention to the faults, and record in Table 1-13 after analysis and summary.

1.3.5 Task Evaluation

The scoring table is shown in Table 1-12.

Table 1-12 Scoring table

Task content	Assessment requirements	Scoring criteria	Allotment	Points deducted	Score
Preparation	(1) Find out the fault of the required components to ensure the normal use of each component; (2) Draw schematic diagram	Preparatory work is the entry stage of the whole implementation process. Only after this stage is completed, can the next step be carried out. If this step is not done or fails, the whole project will be scored 0 point	10		

Continued Table 1-12

Task content	Assessment requirements	Scoring criteria	Allotment	Points deducted	Score
Description of working principle	Be able to correctly explain the working principle of ladder diagram	10 points will be deducted for each output function with a mistake	10		
Wiring of PLC system	Complete the hardware wiring correctly	(1) 5 points will be deducted for each wrong line; (2) Points will be deducted according to other circumstances	20		
PLC program input and download	Input, download and monitor the PLC program according to the task requirements	(1) 15 points will be deducted if the software is not used skillfully and the input operation of the program cannot be completed; (2) 5 points will be deducted if the IP address of the program is not set or downloaded; (3) 5 points will be deducted if the operation of botu software monitoring program will not be operated	20		
PLC control system operation demonstration	Correct program, and explain the combine program and hardware	(1) If the output cannot be started, 10 points will be deducted for each place; (2) 5 points will be deducted for each place where the operation phenomenon cannot be analyzed and explained	30		
Safety and civilization production	Comply with 8S management system and safety management system	(1) 10 points will be deducted if no labor protection articles are worn; (2) If there is potential safety hazard in operation, 5 points will be deducted each time until all points are deducted; (3) If the operation site is not cleaned up in time, 2 points will be deducted for each time until all points are deducted	10		
		Total score			

1.3.6 Task Summary

The task analysis summary record is shown in Table 1-13.

Table 1-13 Task analysis summary record

		Task analysis summary record
Fault 1	Fault phenomenon	
	Cause of failure	
	Exclusion process	
Fault 2	Fault phenomenon	
	Cause of failure	
	Exclusion process	
Fault 3	Fault phenomenon	
	Cause of failure	
	Exclusion process	
	Summary	

1.3.7 Task Development

Task requirements: On XK-SX2C advanced maintenance electrician training platform, connect PLC according to the input point in Figure 1-28, and download the program. On the basis of correct connection, turn on the power, operate the input button switch, and observe the operation status of PLC.

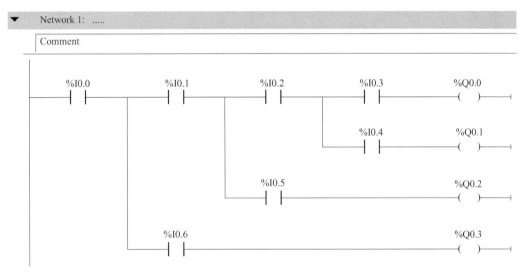

Figure 1-28 Task development procedure

1.3.8 Project Summary

This project consists of three tasks, which is the basis of in-depth study of electrical control and PLC. This knowledge and skill must be firmly established. Task 1.1 mainly introduces the functions, working principles and graphic symbols of commonly used low-voltage circuit breakers, fuses, contactors and switch buttons; Task 1.2 introduces the generation, definition, characteristics and development of PLC, and also focuses on the relevant contents of 1200 series PLC, which belongs to the contents of memory; Task 1.3 introduces the usage of software and the test methods of programs, and pays attention to hardware configuration and the matching of IP address. The external wiring diagram of PLC also needs to be mastered.

Exercises

(1) Briefly describe the development process of PLC.
(2) Briefly describe the advantages of TIA software.
(3) Consult the data, analyze and compare the parameters of 1200 series PLC.
(4) Briefly describe the external wiring mode of Siemens 1215C.
(5) Input the following program (shown in Figure 1-29), and run PLC to observe the operation status.

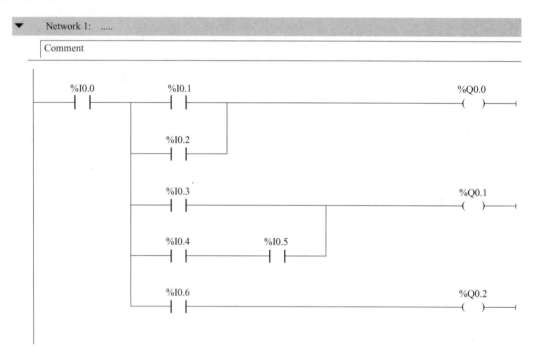

Figure 1-29 Program for the example

Project 2 Instantaneous Start Stop Control of Three-phase Asynchronous Motor

The three-phase asynchronous motor, which has the characteristics of simple structure, convenient manufacture, good running performance, saving various materials and low price, makes it widely used in production and life. For example, lathe, blower, crane, CNC machine tool, elevator, electric trailer, water pump, etc. are inseparable from it.

The control of three-phase asynchronous motor can be divided into instantaneous control and delay control. In this project, two different methods are used to control the three-phase asynchronous motor. One is to use the traditional relay system for control; The other is to use the 1200 series PLC produced by German Siemens company for control. The project is divided into two tasks: the inching self-locking control of asynchronous motor and the forward and reverse control of three-phase asynchronous motor. The use of relay control system and PLC control system is gradually mastered from shallow to deep.

Task 2.1 Inching Self-locking Control of Three-phase Asynchronous Motor

2.1.1 Task Description

The task is divided into two parts: inching control and self-locking control of three-phase asynchronous motor. Each part needs two methods to achieve the same control requirements, that is, using relay control system and PLC control system to complete the inching control and self-locking control of three-phase asynchronous motor.

In the relay control system, the drawing of circuit diagram, the connection of electrical components, the detection before power on, and the analysis after power on are completed. In the PLC control system, in addition to the above requirements, the external wiring of PLC and the programming and debugging of PLC program are also required. Finally, the process of task implementation is summarized and recorded in the corresponding table.

2.1.2 Task Target

(1) Master the wiring, checking and operation of three-phase asynchronous motor lines.

(2) Master the principle of three-phase asynchronous motor inching and self-locking control circuit.

(3) Master the principles of short circuit protection, overload protection and undervoltage pro-

tection.

(4) Understand the connection method of intermediate relay contact and its function.

(5) Understand the differences between PLC control inching and self-locking.

(6) Ability to find, analyze and troubleshoot.

(7) Train students' sense of safe operation and team work.

2.1.3 Task-related Knowledge

2.1.3.1 Thermal Relay

During the operation of the motor, if the load is too large, the current of the motor will exceed its rated value. If the duration is long, the temperature rise of the motor will exceed the allowable temperature rise, which will damage the insulation of the motor and even burn the motor. Therefore, it is necessary to take protective measures for motor overload. When the motor is overloaded, the fuse generally will not fuse, because the fuse connected to the main circuit of the motor is mainly used for the short-circuit protection of the motor, and the allowable current value of the fuse is several times of the rated current of the motor. If the fuse capacity is small, the fuse will often be burnt out when the motor starts. Therefore, motor overload protection needs to take other measures, and the most commonly used is to use thermal relay for overload protection. The actual drawing and electrical symbols are shown in Figures 2-1 and 2-2 respectively.

Figure 2-1 Physical diagram of thermal relay

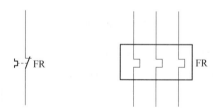

Figure 2-2 Electrical symbols of thermal relay

Working Principle of Thermal Relay

Thermal relay is used for motor or other electrical equipment, electrical circuit overload protection of electrical appliances.

In the actual operation of the motor, if the abnormal condition of the machinery or the abnormal circuit causes the motor to encounter overload in the process of driving the production machinery, the motor speed will decrease, the current in the winding will increase, and the winding temperature of the motor will increase. If the overload current is small and the overload time is short, and the motor winding does not exceed the allowable temperature rise, this overload is allowed. However, if the overload time is long and the overload current is large, the temperature rise of the motor winding will exceed the allowable value, aging the motor winding, shortening the service life of the motor, and even burning the motor winding in serious cases. Therefore, this kind of overload can not be borne by the motor. Thermal relay is to cut off the motor circuit in case of overload that the motor can't bear by using the thermal effect principle of current to provide overload protection for the motor.

When the thermal relay is used for overload protection of the motor, the thermal element is connected in series with the stator winding of the motor, the normally closed contact of the thermal relay is connected in series with the control circuit of the electromagnetic coil of the AC contactor, and the setting current adjusting knob is adjusted to make the herringbone paddle and the push rod at a proper distance. When the motor works normally, the current passing through the thermal element is the rated current of the motor. The thermal element is heated and the bimetallic sheet is bent after heated, so that the push rod just contacts the herringbone paddle, but the herringbone paddle cannot be pushed. The normally closed contact is in the closed state, the AC contactor remains closed, and the motor operates normally.

If the motor is overloaded, the current in the winding will increase. Through the increase of the current in the thermal relay element, the bimetal temperature will rise higher, the bending degree will increase, and the herringbone paddle will push the normally closed contact, so that the contact will be disconnected and the coil circuit of the AC contactor will be disconnected. The contactor will release and cut off the power supply of the motor, and the motor will be protected when it stops.

The working principle of the thermal relay is that the current flowing into the thermal element generates heat, which makes the bimetallic sheet with different expansion coefficient deform. When the deformation reaches a certain distance, it will push the connecting rod to act, so that the control circuit is disconnected, the contactor loses power, the main circuit is disconnected, and the overload protection of the motor is realized. As the overload protection element of motor, thermal relay is widely used in production because of its small volume, simple structure and low cost.

Daily Maintenance of Thermal Relay

The daily maintenance of thermal relay are as follows:

(1) The reset of thermal relay shall take a certain time after action, the automatic reset time shall be completed within 5min, and the manual reset can only be pressed after 2min.

(2) In case of short circuit fault, the main goal is to check whether the thermal element and bimetallic sheet are deformed. If there is any abnormal condition, adjust it in time, but do not remove the element.

(3) The thermal relay in use shall be inspected once a week. The specific contents are: Whether the thermal relay is overheated, peculiar smell and discharged; Whether the screws of all parts are loose, fallen off and removed poorly; Whether the surface is damaged and clean.

(4) The thermal relay in use shall be maintained once a year. The specific contents are: cleaning, checking and repairing parts, testing the insulation resistance to be greater than $1M\Omega$, and powering on for verification. For the calibrated thermal relay, except for the wiring screws, other

screws shall not be moved casually.

(5) When replacing the thermal relay, the newly installed thermal relay must meet the original specifications and requirements.

(6) The wiring is checked regularly for looseness, and the bimetallic sheet is never bent during maintenance.

Practices

(1) Observe the appearance of thermal relay, and check the label and technical parameters.

(2) Think about how to connect wires, and write the corresponding wire number and function of terminals in Table 2-1.

Table 2-1 Function table of thermal relay terminal

No.	Terminal name	Function	No.	Terminal name	Function
1			5		
2			6		
3			7		
4			8		

2.1.3.2 Analysis of Inching and Self-locking Control Principle

The principle is analyzed and introduced in four parts: the inching circuit controlled by relay system, the inching circuit controlled by PLC system, the self-locking circuit controlled by relay system, and the self-locking circuit controlled by PLC system.

Inching Circuit Controlled by Relay System

Inching control circuit is the simplest control circuit to control motor with button and contactor. Inching control includes: pressing the button, the motor is powered on for operation; releasing the button, the motor is powered off and stops running. This control method is commonly used in the hoist motor of the electric hoist and the motor control of the lathe rapid movement. The control circuit is usually represented by the electrical graphic symbols and text symbols specified in the national standards and drawn into the control circuit schematic diagram. It is drawn according to the physical wiring circuit to express the working principle of the control circuit, and its schematic diagram is shown in Figure 2-3.

The inching control schematic can be divided into two parts: the main circuit and the control circuit. The main circuit is the circuit from the power supply L_1, L_2, and L_3 to the motor M through the main contact of the power switch QS, fuse FU_1, and contactor KM. It flows through a large circuit. The control circuit is composed of fuse FU_2, button SB and the coil of contactor KM, and the current flowing is small.

When the motor needs to be inched, the power switch QS is closed firstly, then the inching button SB is pressed, the contactor coil KM will be electrified, the armature will be closed, driving its three pairs of normally open main contacts KM to close, and the motor M will be connected

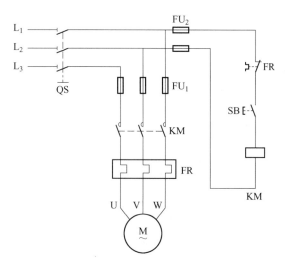

Figure 2-3 Inching circuit diagram of relay system control

to the power supply for starting operation. After the button of SB is released, the contactor coil is de energized, and the armature is reset under the action of spring force, driving its three pairs of normally open main contacts to open. The motor is de energized and stopped.

Working principle of inching control circuit includes: after closing the power switch QS, when starting the motor, the steps of operation are: Pressing SB→KM coil energized→KM main contact closed→the motor M running; When stopping the motor, the steps of operation are: Releasing SB→Turning off KM coil→Turning off KM main contact→Stopping the motor M.

Practice

Observe the circuit diagram as shown in Table 2-3, and

(1) analyze the operation process independently and draw the circuit diagram by hand.

(2) think about how to connect.

Inching Circuit Controlled by PLC System

Many advantages of PLC are introduced in Project 1, which will not be described here. After the introduction of PLC for control, it can enhance the stability of the system, and improve the running speed and other characteristics. Its schematic diagram is shown in Figures 2-4 and 2-5.

It can be seen that, as shown in Figure 2-4, the intermediate relay KA is used as a bridge for weak current control and strong current control, which can protect PLC. Even if some PLC output can be directly connected to AC, but for the sake of safety, intermediate relay is often used in practical application. KA used in this task supplies 24V DC, which includes a coil and a pair of normally open contacts. When SB is pressed, PLC gets input signal, which makes PLC output have signal after internal operation calculation. When KA coil is energized, KA normally open contact is closed, which makes KM coil energized. When armature is closed, three pairs of normally open main contact KM are closed, and the motor M is connected with electric source to start operation. FR plays a protective role and can be considered as an input point in PLC circuit. When

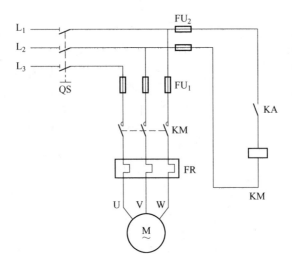

Figure 2-4 Main circuit wiring diagram of inching control (PLC)

SB is released, the input signal of PLC is interrupted, and the output signal of PLC disappears after internal operation calculation. KA coil is de-energized, which restores KA normally open contact to its original state, resulting in KM coil de-energized. Armature is reset under the effect of spring force, which drives its three pairs of normally open main contacts to open, and motor is de energized and stopped.

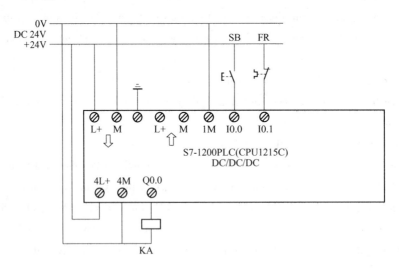

Figure 2-5 Wiring diagram of PLC control circuit of inching control

Practice

Observe the two parts of the circuit diagram, as shown in Figures 2-4 and 2-5, and
   (1) analyze its operation process independently and draw the circuit diagram by hand;
   (2) think about how to connect.

Self Locking Circuit Controlled by Relay System

In order to realize the continuous operation of the motor, the forward rotation control circuit with contactor self-locking can be used. A normally open auxiliary contact of the contactor needs to be connected in parallel at both ends of the start button SB_2. In the control circuit, a stop button SB_1 can be connected in series to stop the motor. The self-locking forward control circuit of the contactor can not only make the motor run continuously, but also has the functions of under voltage protection and loss of voltage (zero voltage) protection. The schematic diagram is shown in Figure 2-6.

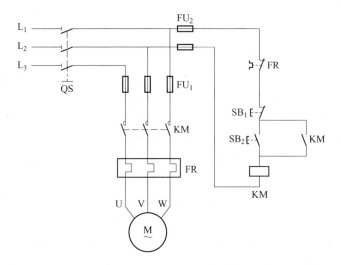

Figure 2-6 Circuit diagram of self-locking control relay

- **Undervoltage Protection**

Undervoltage means that the line voltage is lower than the rated voltage to be applied to the motor. Undervoltage protection refers to the protection that when the line voltage is lower than a certain value, the motor can automatically break away from the power supply voltage and stop running to avoid the motor running under undervoltage. Why do motors have undervoltage protection is that when the power supply voltage drops, the current of the motor will rise, while the motor is running. The more serious the voltage drops, the more serious the current rises. In serious cases, the motor will burn out.

When the motor is running, and the power supply voltage is reduced to a lower level (generally below 85% of the working voltage), the magnetic flux of the contactor coil becomes very weak, and the electromagnetic attraction is insufficient. The moving iron core is released under the action of the reaction spring, the self-locking contact is disconnected and the self-locking is lost. At the same time, the main contact is also disconnected, and the motor stops running, which is protected.

- **Loss of Voltage (Zero Pressure) Protection**

Loss of voltage protection means that when the power supply is temporarily cut off due to some external reasons during the operation of the motor, the power supply of the motor can be cut off automatically. When the power supply is restored, the motor can not be started by itself. If no preventive measures are taken, it is easy to cause personal accidents. The forward rotation control circuit with contactor self-locking is adopted. Since the self-locking contact and the main contact are disconnected together when the power is cut off, neither the control circuit nor the main circuit will be connected by themselves. Therefore, if the button is not pressed when the power supply is restored, the motor will not start by itself.

- **Overload Protection**

Overload protection is a kind of protection that can automatically cut off the power supply of the motor and make the motor stop running when the motor is overloaded. The most commonly used is to use thermal relay for overload protection. During the operation of the motor, such as long-term overload, frequent operation or open phase operation, the current of the motor stator winding may exceed its rated value, but the current does not reach the fuse, which will cause the overheating temperature of the motor stator winding to rise. If the temperature exceeds the allowable temperature rise, the insulation will be damaged. The service life of the motor will be greatly shortened, and even the motor will be burnt in serious cases. Therefore, overload protection measures must be taken for the motor.

During the operation of the motor, the current exceeds the rated value due to overload or other reasons. After a certain period of time, the thermal elements of the thermal relay FR connected in series in the main circuit are bent by heating. Through the action mechanism, FR normally closed contact connected in series in the control circuit is disconnected, the control circuit is cut off, the coil of the contactor km is cut off, the main contact is disconnected, and the motor M stops running, reaching the over limit, for the purpose of protection.

Practice

Observe the circuit diagram of self-locking control relay, and

(1) analyze its operation process independently, and draw the circuit diagram by hand;

(2) think about how to connect.

- **Self Locking Circuit Controlled by PLC System**

The schematic diagram of self-locking circuit controlled by PLC system is shown in Figures 2-7 and 2-8.

It can be seen that, as shown in Figure 2-7, the intermediate relay KA is used as a bridge for weak current control and strong current control. When SB_1 is pressed, PLC gets the input signal, which makes the output of PLC have signal after internal operation calculation. The coil of KA is energized, which makes the normally open contact of KA close, and makes the coil of KM energized. The armature is closed, which drives its three pairs of normally open main contact KM to close, and the motor M is connected to the electric source to start operation. FR plays a protective role and it can be considered as an input point in PLC circuit.

When SB_2 is pressed, the input signal of PLC is interrupted, and the output signal of PLC disappears after internal operation calculation. The coil of KA is de-energized, which restores the

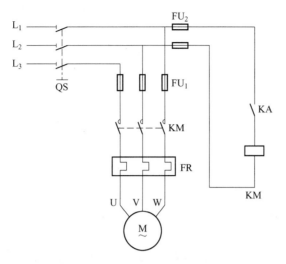

Figure 2-7 Self locking control main circuit wiring diagram (PLC)

normally open contact of KA to the original state, resulting in the coil de-energized of KM. The armature is reset by the spring force, which drives its three pairs of normally open main contacts to open, and the motor is de-energized and stopped.

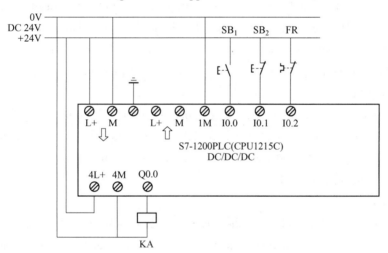

Figure 2-8 Wiring diagram of self-locking control PLC control circuit

Practice

Observe the two parts of the circuit diagram as shown in Figures 2-7 and 2-8, and
    (1) analyze its operation process independently and draw the circuit diagram by hand;
    (2) think about how to connect.

2.1.3.3　Cognition of Input and Output Instructions

Input instructions are mostly used to represent the state of contact, including normally open contact and normally closed contact. Input instructions are mostly used to represent the state of common coil, including power on state and power off state. They are the most basic instructions in PLC programming, and also the entry conditions for learning PLC programming well. The example program as shown in Figure 2-9 is illustrated as follows.

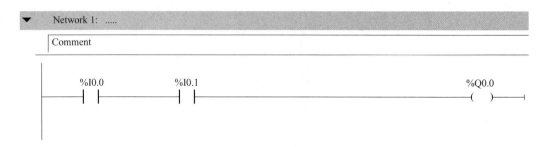

Figure 2-9 Example program of input and output instructions

Input Command

The input command is used to input the state of the contact. The normally open and normally closed contact commands are represented by the contacts in the ladder diagram (LAD). As shown in Figure 2-9, both I0.0 and I0.1 are input commands, indicating normally open contacts. In ladder diagram (LAD), input contact commands refer to the connection between the contacts and the left bus. When the contact is normally closed, the energy flow can pass through the contact; When the contact is normally open, the energy flow can not pass through.

Output Instruction

The output instruction is to write the logical operation result into the output image register, and finally determine the output terminal status. In LAD, the output command is represented as a coil. As shown in Figure 2-9, Q0.0 is the output instruction, which is a common coil, indicating the state of power on or power off.

Comprehensive Application

The coil represented by Q0.0 also has the characteristics of KM coil, that is, when the coil is powered on, its corresponding normally open normally closed contact acts; When the coil is powered off, its corresponding normally open normally closed contact restores to the original state. As shown in Figure 2-10, when the front three normally open contacts are closed, the energy flow reaches the coil and energizes the coil Q0.0. Because the coil Q0.0 is energized, the normally open contact of Q0.0 acts and the normally open state changes to the normally closed state. In this way, even if I0.0 is disconnected, Q0.0 will not lose power. Unless I0.1 or I0.2 is disconnected, Q0.0 can be powered off.

Practice

Analyze Figure 2-10 to explain why Q0.0 will not lose power even if I0.0 is disconnected after Q0.0 is powered on.

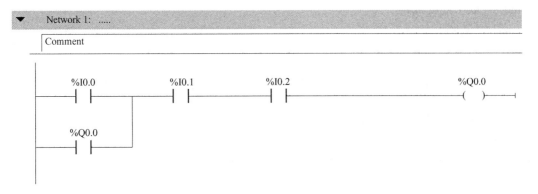

Figure 2-10 Example program of integrated application

2.1.4 Task Implementation

On the XK-SX2C advanced maintenance electrician training platform, the following four tasks are completed: the control of relay system for three-phase asynchronous motor inching, the control of PLC system for three-phase asynchronous motor inching, the control of relay system for three-phase asynchronous motor self-locking, and the control of PLC system for three-phase asynchronous motor self-locking. Each task is required to be powered on the basis of making sure the connection is correct. In the process of implementation, if there is a fault, it is necessary to find, analyze and eliminate the fault. Pay attention to the cultivation of professional core literacy such as safety awareness and team awareness.

2.1.4.1 Relay Systems Inching Control of Three-phase Asynchronous Motor

Task Requirements

Connect the line according to Figure 2-3, complete the inching control of the three-phase asynchronous motor, and then score the project after completion, as shown in Table 2-4.

Operation Steps

The operation steps are as follows:

(1) Find out the fault of the required components to ensure the normal use of each component.

(2) Draw the schematic diagram and mark the line number.

(3) Connect the lines according to the prepared drawings.

(4) Use the multimeter to check before power on. In case of any fault, remove it in time and record the summary.

(5) Power on and observe the operation status of the system. In case of any fault, remove it in time and record the summary.

(6) Connect the three-phase asynchronous motor into the line, and observe its operation state with power on.

(7) Summarize and record.

Analysis and Summary

Summarize the whole task implementation process, find out the deficiencies, especially pay attention to the faults, and record in Table 2-5 after analysis and summary.

2.1.4.2 Control of PLC System to Three-phase Asynchronous Motor Inching

Task Requirements

According to Figure 2-4 and Figure 2-5 the circuit is connected to complete the inching control of three-phase asynchronous motor. Finally, score the project after completion, as shown in Table 2-4.

Operation Steps

The operation steps are as follows:

(1) Find out the fault of the required components to ensure the normal use of each component.

(2) Draw the schematic diagram and mark the line number.

(3) Connect the power line of PLC and measure it with multimeter to ensure no error before power on.

(4) Determine I/O address assignment and fill in Table 2-2.

Table 2-2 I/O address allocation

Input signal				Output signal			
No.	Function	Element	Address	No.	Control object	Element	Address
1				1			
2				2			
3				3			

(5) According to the I/O address assignment, complete the line connection of the input terminal.

(6) Power on PLC after checking.

(7) Write program, and download to PLC for debugging.

(8) Complete the connection of the whole line according to the prepared drawings.

(9) Use the multimeter to check before power on. In case of any fault, remove it in time and record the summary.

(10) Power on and observe the operation status of the system. In case of any fault, remove it in time and record the summary.

(11) Connect the three-phase asynchronous motor into the line, and observe its operation state with power on.

(12) Summarize and record.

Analysis and Summary

Summarize the whole task implementation process, find out the deficiencies, especially pay attention to the faults, and record in Table 2-5 after analysis and summary.

2.1.4.3 Self-locking Control of Three-phase Asynchronous Motor by Relay System

Task Requirements

The lines are connected according to Figure 2-6 to complete the self-locking control of the three-phase asynchronous motor, and then score the project after completion, as shown in Table 2-4.

Operation Steps

The operation steps are as follows:

(1) Find out the fault of the required components to ensure the normal use of each component.

(2) Draw the schematic diagram and mark the line number.

(3) Connect the lines according to the prepared drawings.

(4) Use the multimeter to check before power on. In case of any fault, remove it in time and record the summary.

(5) Power on and observe the operation status of the system. In case of any fault, remove it in time and record the summary.

(6) Connect the three-phase asynchronous motor into the line, and observe its operation state with power on.

(7) Summarize and record.

Analysis and Summary

Summarize the whole task implementation process, find out the deficiencies, especially pay attention to the faults, and record in Table 2-5 after analysis and summary.

2.1.4.4 Self-locking Control of Three-phase Asynchronous Motor by Relay System

Task Requirements

The lines are connected according to Figures 2-7 and 2-8, the self-locking control of the three-phase asynchronous motor is completed, and the project is scored after completion, as shown in Table 2-4.

Operation Steps

The operation steps are as follows:

(1) Find out the fault of the required components to ensure the normal use of each component.

(2) Draw the schematic diagram and mark the line number.

(3) Connect the power line of PLC and measure it with multimeter to ensure no error before power on.

(4) Determine I/O address assignment and fill in Table 2-3.

Table 2-3 I/O address allocation

Input signal				Output signal			
No.	Function	Element	Address	No.	Control object	Element	Address
1				1			
2				2			
3				3			

(5) According to the I/O address assignment, the line connection of the input terminal is completed.

(6) Power on PLC after checking.

(7) Write program, and download to PLC for debugging.

(8) Complete the connection of the whole line according to the prepared drawings.

(9) Use the multimeter to check before power on. In case of any fault, remove it in time and record the summary.

(10) Power on and observe the operation status of the system. In case of any fault, remove it in time and record the summary.

(11) Connect the three-phase asynchronous motor into the line, and observe its operation state with power on.

(12) Summarize and record.

Analysis and Summary

Summarize the whole task implementation process, find out the deficiencies, especially pay attention to the faults, and record in Table 2-5 after analysis and summary.

2.1.5 Task Evaluation

The fourth task relay system is evaluated for the self-locking control of three-phase asynchronous motor, and the scoring results are filled in Table 2-4.

Table 2-4 Scoring table

Task content	Assessment requirements	Scoring criteria	Allotment	Points deducted	Score
Preparation	(1) Find out the fault of the required components to ensure the normal use of each component; (2) Draw the schematic diagram and mark the line number	Preparatory work is the entry stage of the whole implementation process. Only after this stage is completed, can the next step be carried out. If this step is not done or fails, the whole project will be scored 0 point	10		
Description of working principle	Can correctly explain the working principle of the whole system	(1) 5 points will be plused for the main circuit function description; (2) 5 points will be plused for PLC line function description	10		
PLC control system main control circuit wiring	Correctly complete the wiring of PLC main circuit and control circuit	(1) 5 points will be deducted for each wrong line; (2) Points will be deducted according to other circumstances	20		

Continued Table 2-4

Task content	Assessment requirements	Scoring criteria	Allotment	Points deducted	Score
PLC programming and downloading	Input, download and monitor the PLC program according to the task requirements	(1) 15 points will be deducted if the software is not used skillfully and the input operation of the program cannot be completed; (2) 5 points will be deducted if the program's IP address is not set and downloaded; (3) 5 points will be deducted if the PLC's stop and operation status cannot be controlled by Bertrand software; (4) 5 points will be deducted if the programs operation is not controlled by Bertrand software	20		
PLC control system operation demonstration	Correctly demonstrate the start and stop of the motor, and can be explained in combination with the program and hardware	(1) 10 points will be deducted if the circuit fails to start; (2) 10 points will be deducted if the circuit fails to stop; (3) 5 points will be deducted for each place where the operation phenomenon cannot be analyzed and explained	30		
Safety and civilization production	Comply with 8S management system and safety management system	(1) 10 points will be deducted if no labor protection articles are worn; (2) If there is potential safety hazard in operation, 5 points will be deducted each time until all points are deducted; (3) If the operation site is not cleaned up in time, 2 points will be deducted for each time until all points are deducted	10		
		Total score			

2.1.6 Task Summary

The task analysis summary record is shown in Table 2-5.

Table 2-5 Task analysis summary record

		Task analysis summary record
Fault 1	Fault phenomenon	
	Cause of failure	
	Exclusion process	

Continued Table 2-5

		Task analysis summary record
Fault 2	Fault phenomenon	
	Cause of failure	
	Exclusion process	
Fault 3	Fault phenomenon	
	Cause of failure	
	Exclusion process	
	Summary	

2.1.7 Task Development

Task requirements: Two control systems are respectively used to realize the sequential start and stop control of two three-phase asynchronous motors.

Requirements: Both motors have their own start and stop buttons. After the first motor is started, the second motor can be started. If both motors are in operation state, the first motor can be stopped after the second motor is stopped.

Task 2.2 Forward and Reverse Control of Three-phase Asynchronous Motor

2.2.1 Task Description

This task is the forward and reverse control of three-phase asynchronous motor, which requires two methods to achieve the same control requirements, that is, using relay control system and PLC control system to complete the forward and reverse control of three-phase asynchronous motor.

In the relay control system, the drawing of circuit diagram, the connection of electrical components, the detection before power on, and the analysis after power on are completed. In the PLC control system, the external wiring of PLC and the programming and debugging of PLC program are also required, in addition to the above requirements. Finally, the process of task implementation is summarized and recorded in the corresponding table.

2.2.2 Task Target

(1) Understand the meaning, function and implementation of interlock.

(2) Learn to realize various methods and precautions of motor forward and reverse.

(3) Master the principle of forward and reverse circuit of three-phase asynchronous motor.

(4) Understand the differences between contactor interlock forward and reverse and double interlock forward and reverse.

(5) Ability to find, analyze and troubleshoot.

(6) Train students' sense of safe operation and team work.

2.2.3 Task-related Knowledge

2.2.3.1 Basic Knowledge of Motor Forward and Reverse

The forward and reverse rotation of the motor represents the clockwise and counter clockwise rotation of the motor. Clockwise rotation of the motor means positive rotation of the motor, and anticlockwise rotation means reverse rotation of the motor. Forward and reverse control circuit diagram and its principle analysis in order to realize the forward and reverse of the motor, it is only necessary to connect any two relative adjustment wires in the three-phase power supply incoming line of the motor to achieve the purpose of reverse. The forward and reverse rotation of the motor are widely used, such as the electric planer, bench drill, wire cutter, dryer and lathe for driving and woodworking.

At first, people need to reverse some kind of equipment, but this method is cumbersome in practical use. With the birth of the contactor, the forward and reverse circuit of the motor also has further development. It can control the forward and reverse of the motor more flexibly and conveniently, and add the protection circuit-interlock and double interlock in the circuit. It can realize low voltage and long-distance frequent control.

In order to enable the motor to rotate forward and reverse, two contactors KM_1 and KM_2 can be used to change the phase sequence of the three-phase power supply of the motor, but the two contactors cannot be closed at the same time. If the two contactors are closed at the same time, it will cause a short-circuit accident of the power supply. In order to prevent this kind of accident, reliable interlock shall be adopted in the circuit, and the electric motor with button interlock, contactor interlock, button and contactor double interlock shall be used. The control circuit for forward and reverse operation of the unit.

2.2.3.2 Analysis of Control Principle of Forward and Reverse Circuit

Contactor Interlock Positive and Reverse Control Circuit

As shown in Figure 2-11, two contactors are used in the line, i.e. forward transfer contact KM_1 and reverse contact KM_2, which are respectively controlled by forward transfer the button SB_2 and reverse the button SB_3. When KM_1 main contact is connected, three-phase power supply L_1, L_2, and L_3, and U—V—W phase sequence is connected to the motor. When the main contact of KM_2 is connected, the three-phase power supply L_1, L_2, and L_3 are connected to the motor according to W—V—U phase sequence, i.e. the phase sequence of W and U is reversed. Correspondingly, there are two control circuits: One is the forward control circuit composed of forward button SB_2 and KM_1 coil; The other is the reverse control circuit composed of reverse button SB_3 and KM_2 coil, etc. So when two contactors work separately, the rotation direction of the motor is opposite. It must be pointed out that the line requires that the contactors KM_1 and KM_2 cannot be connected at the same time. Otherwise, their main contacts will be closed at the same time, which will cause the short circuit of L_1, L_2, L_3 and two-phase power supplies. For this reason, a pair of normally closed auxiliary contacts are connected in series with each other in the branches of the coils of contactors KM_1 and KM_2. The normally closed auxiliary contacts of reversing contactor KM_2 are connected in series in the forward control circuit and the normally closed auxiliary contacts of reversing contactor KM_1 are connected in series in the reverse control circuit. To ensure that contactors KM_1 and KM_2 are not energized at the same time. These two pairs of normally closed auxiliary contacts of KM_1 and KM_2 are called interlocking (or interlocking) in the circuit, and these two pairs of contacts are called interlocking contacts (or interlocking contacts).

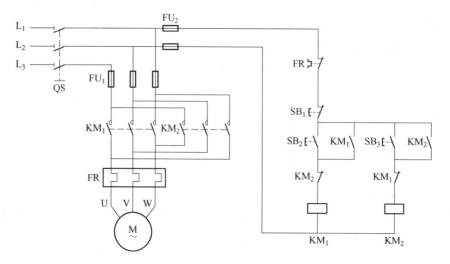

Figure 2-11 Control circuit diagram of contactor interlock forward and reverse

Figure 2-11 is a typical circuit of motor forward and reverse control, but to change the direction of the motor in this circuit, press the stop button SB_1 firstly, and then press the reverse button SB_3, so as to make the motor reverse and inconvenient operation.

Button Interlock Forward and Reverse Control Circuit

In order to overcome the disadvantage of inconvenient operation of forward and reverse control circuit of contactor interlocking, the forward button SB_2 and reverse button SB_3 are replaced by two composite buttons, and the normally closed contacts of two composite buttons replace the interlocking contacts of contactor, thus forming the forward and reverse control circuit of button interlocking, as shown in Figure 2-12.

The working principle of the control circuit is basically the same as that of the forward and reverse control circuit interlocked by the contactor, except that when the motor changes from positive to reverse, it can be realized by directly pressing the reverse button SB_3, without pressing the stop button SB_1 firstly. This not only ensures that the coils of KM_1 and KM_2 will not be powered on at the same time, but also directly press the reverse button to realize the reverse without pressing the stop button. Similarly, if the motor changes from reverse operation to forward op-

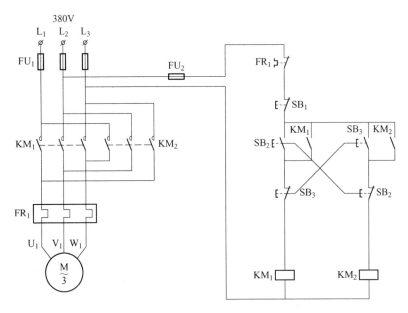

Figure 2-12 Forward and reverse control circuit diagram of button interlock

eration, just press forward button SB_2 directly.

The advantage of this circuit is convenient operation. The disadvantage is that it is easy to produce two-phase short circuit fault. For example, when the main contact of forward contact KM_1 is fused or stuck by sundries, the main contact cannot be disconnected even if the coil of KM_1 loses power. If the reverse button SB_2 is directly pressed, KM_2 will be powered on and the contact will be closed, which will inevitably cause two-phase short circuit fault of power supply. Therefore, it is not safe to use this line. In practice, the forward and reverse control circuit with double interlock of button and contactor is often used.

Forward and Reverse Control Circuit with Double Interlocking of Button and Contactor

In order to overcome the shortcomings of contactor interlocking forward and reverse control circuit and button interlocking forward and reverse control circuit, contactor interlocking is added on the basis of button interlocking to form double interlocking forward and reverse control circuit of button and contactor, as shown in Figure 2-13. The line has the advantages of two kinds of interlocking control lines, convenient operation, and reliable operation.

Practices

(1) This paper analyzes and discusses the principle of the above three kinds of forward and reverse control lines, and explains why KM_1 and KM_2 can't be attracted at the same time.

(2) Draw the above three kinds of control circuit diagram, mark the line number, and think about how to connect.

2.2.3.3 Analysis of Forward and Reverse Circuit Controlled by PLC System

Due to the shortcomings of the button interlock forward and reverse control circuit, it is not used

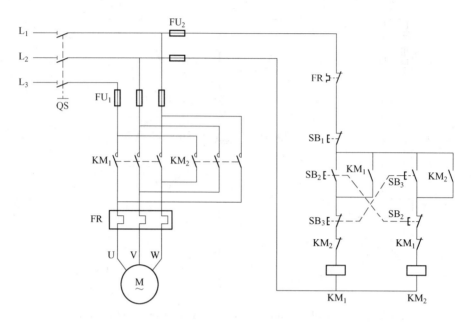

Figure 2-13 Forward and reverse control circuit diagram of double interlocking of button and contactor

basically, so only the remaining two forward and reverse control modes are introduced. The addition of PLC greatly simplifies the control circuit, making the hardware wiring diagram of contactor interlock control circuit and double interlock control circuit the same as shown in Figures 2-14 and 2-15. The only difference is the difference of program.

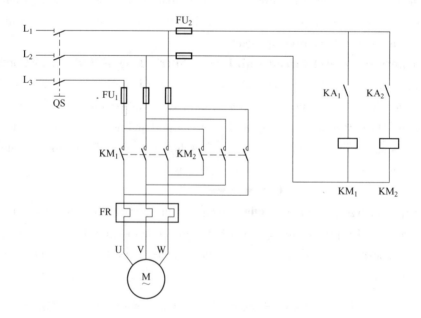

Figure 2-14 Forward and reverse main circuit wiring diagram (PLC)

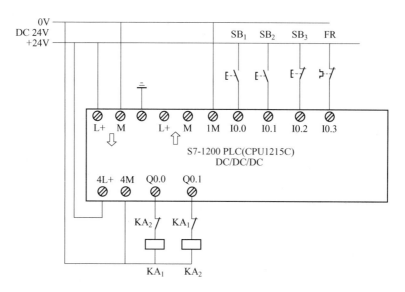

Figure 2-15　Forward and reverse PLC control circuit wiring diagram

Circuit Diagram

The intermediate relay KA is used as a bridge for weak current control and strong current control. It can protect PLC. Even if some PLC output can be directly connected to AC, but for the sake of safety, intermediate relay is often used in practical application. When SB_1 is pressed, PLC gets the input signal, which makes the output of PLC have signal after internal operation calculation. The coil where KA_1 is located is electrified, which makes the normally open contact of KA_1 close, causing the coil of KM_1 to be electrified. The armature is closed, driving its three pairs of normally open main contact KM_1 to close, and the motor is in transit. The function of SB_2 is similar to that of SB_1, which can be analyzed by readers themselves. FR plays a protective role and it can be considered as an input point in PLC circuit. SB_3 is the stop button of the system.

Compilation of Interlock Program

The reference program is shown in Figure 2-16 and Figure 2-17, which is the program of contactor interlock forward and reverse and double interlock forward and reverse respectively.

Practices

(1) Analyze Figure 2-14 and Figure 2-15, and

1) independently analyze its operation process, and draw circuit diagram by hand;

2) think about how to connect.

(2) Download the sample program to PLC and observe the running state of PLC.

(3) Think about how to write PLC control button, contactor double interlock positive and reverse circuit program.

2.2.4　Task Implementation

On the XK-SX2C advanced maintenance electrician training platform, the following four tasks are

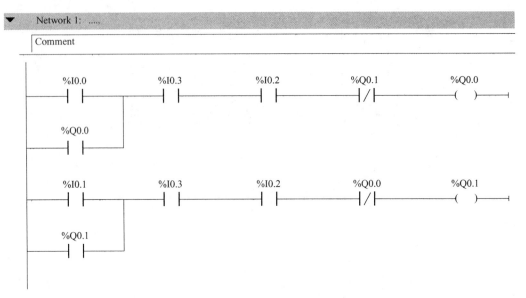

Figure 2-16 Contactor interlock forward and reverse reference procedure

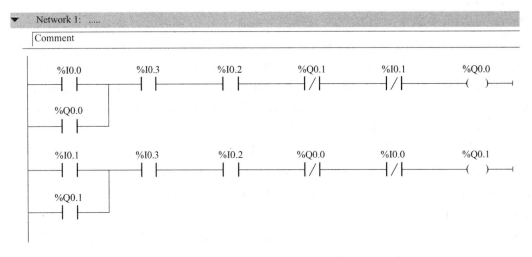

Figure 2-17 Double interlock forward and reverse reference procedure

completed: the control of the relay system for the forward and reverse interlocking of the three-phase asynchronous motor contactor, the control of the PLC system for the forward and reverse interlocking of the three-phase asynchronous motor contactor, the control of the relay system for the forward and reverse interlocking of the three-phase asynchronous motor double, and the control of the PLC system for the forward and reverse interlocking of the three-phase asynchronous motor double. Each task is required to be powered on the basis of making sure the connection is correct. In the process of implementation, if there is a fault, it is necessary to find, analyze and eliminate the fault. Pay attention to the cultivation of professional core literacy such as safety awareness and team awareness.

2.2.4.1 Control of Relay System to Forward and Feverse of Contactor Interlock of Three-phase Asynchronous Motor

Task Requirements

Connect the circuit according to Figures 2-14 and 2-15, complete the forward and reverse interlock control of the contactor of the three-phase asynchronous motor, and then score the project after completion, as shown in Table 2-8.

Operation Steps

The operation steps are as follows:

(1) Find out the fault of the required components to ensure the normal use of each component.

(2) Draw the schematic diagram and mark the line number.

(3) Connect the lines according to the prepared drawings.

(4) Use the multimeter to check before power on. In case of any fault, remove it in time and record the summary.

(5) Power on and observe the operation status of the system. In case of any fault, remove it in time and record the summary.

(6) Connect the three-phase asynchronous motor into the line, and observe its operation state with power on.

(7) Summarize and record.

Analysis and Summary

Summarize the whole task implementation process, find out the deficiencies, especially pay attention to the faults, and record in Table 2-9 after analysis and summary.

2.2.4.2 Control of PLC System for Forward and Reverse Interlock of Three-phase Asynchronous Motor Contactor

Task Requirements

Connect the circuit according to Figures 2-14 and 2-15, complete the forward and reverse interlock control of the contactor of the three-phase asynchronous motor, and the score the project after completion, as shown in Table 2-8.

Operation Steps

The operation steps are as follows:

(1) Find out the fault of the required components to ensure the normal use of each component.

(2) Draw the schematic diagram and mark the line number.

(3) Connect the power line of PLC and measure it with multimeter to ensure no error before power on.

(4) Determine I/O address assignment and fill in Table 2-6.

Table 2-6 I/O address allocation

	Input signal				Output signal		
No.	Function	Element	Address	No.	Control object	Element	Address
1				1			
2				2			
3				3			
4				4			

(5) According to the I/O address assignment, complete the line connection of the input terminal.

(6) Power on PLC after checking.

(7) Write program, and download to PLC for debugging.

(8) Complete the connection of the whole line according to the prepared drawings.

(9) Use the multimeter to check before power on. In case of any fault, remove it in time and record the summary.

(10) Power on and observe the operation status of the system. In case of any fault, remove it in time and record the summary.

(11) Connect the three-phase asynchronous motor into the line, and observe its operation state with power on.

(12) Summarize and record.

Analysis and Summary

Summarize the whole task implementation process, find out the deficiencies, especially pay attention to the faults, and record in Table 2-9 after analysis and summary.

2.2.4.3 Control of Double Interlock Forward and Reverse of Three-phase Asynchronous Motor by Relay System

Task Requirements

Connect the lines according to Figures 2-14 and 2-15, complete the double interlock forward and reverse control of three-phase asynchronous motor, and the score the project after completion, as shown in Table 2-8.

Operation Steps

The operation steps are as follows:

(1) Find out the fault of the required components to ensure the normal use of each component.

(2) Draw the schematic diagram and mark the line number.

(3) Connect the lines according to the prepared drawings.

(4) Use the multimeter to check before power on. In case of any fault, remove it in time and record the summary.

(5) Power on and observe the operation status of the system. In case of any fault, remove it in

time and record the summary.

(6) Connect the three-phase asynchronous motor into the line, and observe its operation state with power on.

(7) Summarize and record.

Analysis and Summary

Summarize the whole task implementation process, find out the deficiencies, especially pay attention to the faults, and record in Table 2-9 after analysis and summary.

2.2.4.4 Control of PLC System to Three-phase Asynchronous Motor Double Interlock Forward and Reverse

Task Requirements

Connect the lines according to Figures 2-14 and 2-15, complete the double interlock forward and reverse control of three-phase asynchronous motor, and score the project after completion, as shown in Table 2-8.

Operation Steps

The operation steps are as follows:

(1) Find out the fault of the required components to ensure the normal use of each component.

(2) Draw the schematic diagram and mark the line number.

(3) Connect the power line of PLC and measure it with multimeter to ensure no error before power on.

(4) Determine I/O address assignment and fill in Table 2-7.

Table 2-7 I/O address allocation

Input signal				Output signal			
No.	Function	Element	Address	No.	Control object	Element	Address
1				1			
2				2			
3				3			
4				4			

(5) According to the I/O address assignment, complete the line connection of the input terminal.

(6) Power on PLC after checking.

(7) Write program, and download to PLC for debugging.

(8) Complete the connection of the whole line according to the prepared drawings.

(9) Use the multimeter to check before power on. In case of any fault, remove it in time and record the summary.

(10) Power on and observe the operation status of the system. In case of any fault, remove it in time and record the summary.

(11) Connect the three-phase asynchronous motor into the line, and observe its operation state with power on.

(12) Summarize and record.

Analysis and Summary

Summarize the whole task implementation process, find out the deficiencies, especially pay attention to the faults, and record in Table 2-9 after analysis and summary.

2.2.5 Task Evaluation

The fourth task relay system is evaluated for the self-locking control of three-phase asynchronous motor, and the scoring results are filled in Table 2-8 respectively.

Table 2-8 Scoring table

Task content	Assessment requirements	Scoring criteria	Allotment	Points deducted	Score
Preparation	(1) Find out the fault of the required components to ensure the normal use of each component; (2) Draw the schematic diagram and mark the line number	Preparatory work is the entry stage of the whole implementation process. Only after this stage is completed, can the next step be carried out. If this step is not done or fails, the whole project will be scored 0 point	10		
Description of working principle	Can correctly explain the working principle of the whole system	(1) 5 points will be plused for the main circuit function description; (2) 5 points will be plused for the PLC line function description	10		
PLC control system main control circuit wiring	Correctly complete the wiring of PLC main circuit and control circuit	(1) 5 points will be deducted for each wrong line; (2) Points will be deducted according to other circumstances	20		
PLC programming and downloading	Input, download and monitor the PLC program according to the task requirements	(1) 15 points will be deducted if the software is not used skillfully and the input operation of the program cannot be completed; (2) 5 points will be deducted if the program IP address is not set and downloaded; (3) 5 points will be deducted if the PLC stop and operation status cannot be controlled by Bertrand software; (4) 5 points will be deducted if the program operation is not controlled by Bertrand software	20		

Continued Table 2-8

Task content	Assessment requirements	Scoring criteria	Allotment	Points deducted	Score
PLC control system operation demonstration	Correctly demonstrate the start, stop and forward and reverse rotation of the motor, and can be explained in combination with the program and hardware	(1) 10 points will be deducted if the circuit fails to start; (2) 10 points will be deducted if the circuit fails to stop; (3) 10 points will be deducted if you can realize the function of forward and reverse; (4) 5 points will be deducted for each place where the operation phenomenon cannot be analyzed and explained	30		
Safety and civilization production	Comply with 8S management system and safety management system	(1) 10 points will be deducted if no labor protection articles are worn; (2) If there is potential safety hazard in operation, 5 points will be deducted each time until all points are deducted; (3) If the operation site is not cleaned up in time, 2 points will be deducted for each time until all points are deducted	10		
		Total score			

2.2.6 Task Summary

The task analysis summary record is shown in Table 2-9.

Table 2-9 Task analysis summary record

Task analysis summary record		
Fault 1	Fault phenomenon	
	Cause of failure	
	Exclusion process	
Fault 2	Fault phenomenon	
	Cause of failure	
	Exclusion process	
Fault 3	Fault phenomenon	
	Cause of failure	
	Exclusion process	
	Summary	

2.2.7 Task Development

Some production machines, such as universal milling machine, require that the worktable can automatically reciprocate and circulate within a certain distance, so that the workpiece can be processed continuously and the production efficiency can be improved. Figure 2-18 shows a schematic diagram of the automatic reciprocating movement of the worktable. The work table is equipped with iron stops 1 and 2, and the machine bed is equipped with stroke switches SQ_1 and SQ_2. When the iron stops collide with the stroke switch, the positive and reverse motor control circuits are automatically changed to make the work table move back and forth automatically. The stroke of the worktable can be adjusted by moving the position of the iron block to adapt to the different requirements of processing parts. SQ_3 and SQ_4 are used for limit protection, that is, limiting position of the workbench. In order to prevent SQ_1 and SQ_2 from failure and the worktable from exceeding the limit position and causing accidents. Please design the schematic diagram.

Figure 2-18 Schematic diagram of automatic reciprocating movement of worktable

Position switch (also known as travel switch, or limit switch) is a kind of control appliance that converts mechanical signal into electrical signal to control the position or travel of moving parts. The position control circuit is to use the iron block on the moving parts to collide with the position switch to make its contact action, so as to connect or disconnect the circuit, to control the mechanical stroke or realize the automatic round-trip of the processing process. The circuit is simple, not affected by various parameters, and only reflects the position of moving parts.

2.2.8 Project Summary

There are two tasks in this project: the inching self-locking control and the forward and reverse control of the three-phase asynchronous motor, which are instantaneous control and have no delay function. This project is the entry-level technology of actual industrial control, which must be mastered. In particular, the ability to understand and apply interlocking knowledge. Interlock can be divided into three modes: button interlock, contactor interlock and double interlock. The first one is low reliability, but it can change the running direction without stopping; The second one is high reliability, but it needs to stop to change the running direction; The third one combines the advantages of the first two, and high reliability, and it does not need to stop to change the running direction at the same time.

Exercises

(1) Draw forward and reverse control circuit diagram of double interlock of button and contactor.

(2) How many kinds of interlock modes are there for three-phase asynchronous motor? And analyze the advantages and disadvantages.

(3) Write a program of three groups of emergency responders. Each group has a emergency answer button and an indicator light. The indicator light of the person who first presses the emergency answer button will be on, and then the indicator light will not respond.

(4) Add a referee on the basis of question (3). The referee has the start and reset buttons in his hands. Only when the referee presses the start button can be rush be answer. The referee presses the reset button and the answer is over.

Project 3　Delay Control of Three-phase Asynchronous Motor

The time-delay control of the motor is very common in our production and life, such as lifting and dropping operation of elevators, opening and closing operation of elevator doors, and starting and stopping of the conveyor conveyor system, etc. The relay control circuit realizes the delay control of the motor through a time relay, and for the PLC control circuit, it implements various delay control of the motor through the timer instruction within the PLC. This project mainly learns how to use the time relay, timer and counter to design the corresponding main circuit and control circuit to control the delayed operation and cyclic operation of the motor. Deeply explain the design of the main circuit and control circuit of the motor delay control.

Task 3.1　Wiring and Debugging of Main Circuit and Control Circuit of Delayed Start of Motor

3.1.1　Task Description

This task uses the relay system and PLC control system to complete the design, wiring and debugging of the three-phase asynchronous motor's delayed start main control circuit.

3.1.2　Task Target

(1) Master the use of time relay.

(2) Master the design, wiring and debugging of the main control circuit of the three-phase asynchronous motor's delayed start relay control system.

(3) Understand the working principle of Siemens S7-1200 PLC timer instruction.

(4) Master the design and wiring of the main control circuit of the three-phase asynchronous motor's delayed start PLC control system.

(5) Independently complete the design and debugging of PLC program.

3.1.3　Task-related Knowledge

3.1.3.1　Cognition of Time Relay

The time relay is an automatic controller that realizes the delayed connection or disconnection of the contacts using the electromagnetic principle or the mechanical action principle. According to the characteristics of contact delay, the time relay can be divided into two types: power-on delay and power-off delay. According to different working principles, time relays can be divided into air damping type, electronic type, digital display type, and electromagnetic type. Among them, electronic time relays are more commonly used. The time relay used in this project is electronic, as shown in Figure 3-1.

Figure 3-1   Physical diagram of electronic time relay

The electrical symbol of electronic time relay is shown in Figure 3-2.

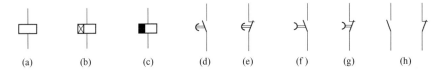

Figure 3-2   Electrical symbol of electronic time relay

As shown in Figure 3-2, (a) is the general form of the time relay coil; (b) is the power-on delay relay coil; (c) is the power-off delay relay coil; (d) and (e) are the delay of the power-on delay relay time contacts; (f) and (g) are the delay contacts of the power-off delay relay; (h) are the instantaneous contacts of the time relay. Whether it is a coil or a contact, the letter symbol of the time relay is KT.

Pull out the time relay from the base, It can be seen that each terminal corresponds to a terminal number. Compare the wiring diagram printed next to the time relay as shown in Figure 3-3. ② and ⑦ are the voltage input terminals. ⑤ and ⑧ are normally closed contacts. ①, ③, ⑧ and ⑥ are normally open contacts. After wiring, insert the time relay into the base.

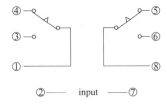

Figure 3-3   Wiring diagram of electronic time relay

Practice

Adjust the time setting of the time relay, and use a multimeter to find the closing and opening contacts of the time relay.

3.1.3.2 Timer Instruction

The S7-1200 series PLC uses IEC standard timer instructions. The number of timers that can be used in the user program is only limited by the CPU memory capacity. Each timer uses a 16-Byte DB structure of the IEC_TIMER data type. Store the timer data specified at the top of the function box or coil instruction. The timer types of S7-1200 series PLC include pulse timer, on-delay timer, off-delay timer, and retentive on-delay timer. Only the pulse timer and the on-delay timer are introduced here.

Pulse Timer

The instruction ladder diagram of the pulse timer is shown in Figure 3-4(a), and its identifier is TP. This instruction is used to survive pulses with a preset width time. The IN pin of the timer instruction is used to enable the timer. The PT pin indicates the timer setting value, Q indicates the timer output state, and ET indicates the timer. The current value of, as shown in Figure 3-4(b), is the instruction format of the pulse timer instruction and the timing chart when the timer instruction is executed.

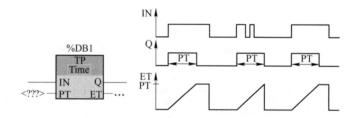

Figure 3-4 Pulse timer symbol and timing diagram

Using the TP instruction, the output of Q can be set to a preset period of time. When the state of the enable end of the timer changes from OFF to ON, the timer instruction can be started and the timer starts counting. No matter how the state of the subsequent enable terminal changes, the output Q is set to a period specified by PT. If the timer is counting, even if the signal of the enable terminal is detected from OFF to ON, the signal state of the output Q will not be affected.

According to the timing diagram of the pulse timer, the following program execution process is analyzed, as shown in Figure 3-5.

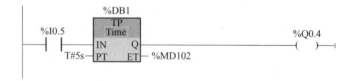

Figure 3-5 Example of pulse timer

When I0.5 is turned ON, the state of Q0.4 is ON. After 5s, the state of Q0.4 becomes OFF. During this 5s, no matter how the state of I0.5 changes, the state of Q0.4 Always remain ON.

On-delay Timer

The instruction identifier of the on-delay timer is TON, as shown in Figure 3-6(a). Turn on the delay timer output terminal Q after the preset delay time, the output state is ON, and the pin definition in the instruction is consistent with the TP timer instruction pin definition. Figure 3-6(b) describes the instruction format and execution timing diagram of the on-delay timer.

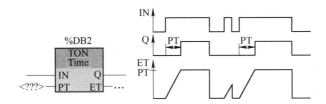

Figure 3-6　Symbols and timing diagram of the pass delay

When the enable terminal IN of the timer is 1, the instruction is started and the timing is started. When the current value ET of the timer is equal to the set value PT, the output Q output is ON. As long as the state of the enable terminal is still ON, the output terminal Q keeps the output ON. If the signal state of the enable terminal becomes OFF, the reset output terminal Q is turned OFF. When the enable terminal turns ON again, the timer function will start again. The working example of the on-delay timer is shown in Figure 3-7.

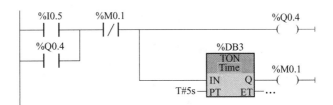

Figure 3-7　Example of on-delay timer

This section of the program mainly completes the procedure of automatically disconnecting after a period of time after starting the output. When I0.5 is turned on, it makes Q0.4 get electric self-locking output which is ON. When Q0.4 output is ON, start the on-delay timer TON to delay the operation of the timer. After a delay of 5s, the output of the timer output Q is in the ON state, making the M0.1 coil energized to ON, so that the program normally closed contact of M0.1 is turned off, and Q0.4 self-lock release is turned off.

Practice

Find the power-on delay timer instruction and pulse timer instruction from the instruction tree of Portal software, and drag them to the programming interface to get familiar.

3.1.4 Task Implementation

3.1.4.1 Wiring and Debugging of Main Control Circuit for Delayed Start of Motor (Relay Control System)

Task Requirements

Jogging the start button, the motor will start with a delay of 10s; Jogging the stop button, the motor will stop immediately. Completing the wiring and debugging of the main circuit and the control circuit, the circuit should have the necessary short-circuit and overload protection measures.

Finally, score the project after completion, as shown in Table 3-2.

Task Analysis

Pressing the start button, the time relay is energized and start timing. To turn off the start button, the time relay is still energized and timed, which requires the help of the self-locking of the intermediate relay. After the timing time expires, the time-delay closing contact of the time relay is closed, so that the contactor coil controls the operation of the motor locks itself, and the motor starts to run. At this time, the tasks of time relay and intermediate relay have been completed, and the circuit where they are located is disconnected through the normally closed contact of the contactor. The main circuit should have a fuse and thermal relay. The control circuit should have a fuse, and a normally closed contact of the thermal relay.

To sum up, the use of time relay to achieve the motor delay start main control circuit is shown in Figure 3-8.

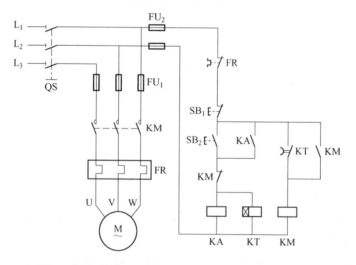

Figure 3-8 Delay start relay main control circuit diagram

Operation Steps

The operation steps are as follows:

(1) On the XK-SX2C advanced maintenance electrician training platform, use jumper wires

to complete the connection between the main circuit and the control circuit, as shown in Figure 3-8.

(2) Set the action time of the time relay to 10s.

(3) Use a multimeter to measure the main circuit and the control circuit separately to check whether the circuit has short circuit, open circuit or wrong connection. If there is a problem, carefully check until the fault is eliminated.

Task Presentation

The task presentation is as follows:

(1) Close the main power supply of the training platform and the circuit isolation switch QS.

(2) Jogging SB_2, the intermediate relay coil pulls in, and the time relay indicator lights. The contactor coil pulls in after 10s, the motor starts, and the intermediate relay and time relay are powered off.

(3) Jogging SB_1, the contactor coil is powered off and released, and the main contact is opened. The motor stops immediately.

(4) Disconnect the isolation switch QS and the main power supply of the training platform.

3.1.4.2 Wiring and Debugging of Main Control Circuit of Delayed Start of Motor (PLC Control System)

Task Requirements

Jogging the start button connected to the PLC input point, the motor will start after a delay of 10s; Jogging the stop button connected to the PLC input point, the motor stops immediately. Completing the wiring and debugging of the main circuit and PLC control circuit, the circuit should have the necessary short-circuit and overload protection measures.

Finally, score the project after completion, as shown in Table 3-2.

Task Analysis

When designing a PLC control system, you should analyze firstly according to the design requirements and determine the PLC I/O address allocation. The PLC control system I/O address allocation table is shown in Table 3-1.

Table 3-1 PLC control system I/O address allocation table

Input signal				Output signal			
No.	Function	Element	Address	No.	Controlled object	Element	Address
1	Start button	SB_1	I0.0	1	Intermediate relay	KA	Q0.0
2	Stop button	SB_2	I0.1	—	—	—	—
3	Overload protection	FR	I0.2	—	—	—	—

Based on the PLC address allocation table, the external wiring diagram of the design PLC is shown in Figure 3-9.

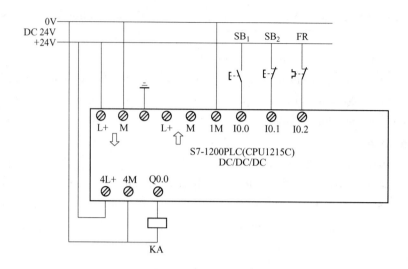

Figure 3-9 Delayed start PLC control circuit wiring diagram

The control of the contactor still needs to control the contactor coil through the normally open contact of the intermediate relay, that is, the weak current is controlled by the strong current, as shown in Figure 3-10.

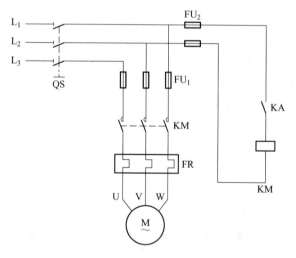

Figure 3-10 Motor main circuit wiring diagram (PLC)

On the basis of the completion of the design of the main circuit and the control circuit, the design of the PLC program should be carried out. The reference program is shown in Figure 3-11.

The program execution and hardware equipment operation are as follows:

(1) Delayed start: Jogging SB_1, I0.0 is turned on for a short time to make the M0.0 coil self-lock, and the IN terminal of the timer is turned on to start timing. When the timer counts to 10s, the Q terminal of the timer is turned on, so that the M0.1 coil is turned on. As the M0.1 coil is connected, the corresponding two contacts are activated.

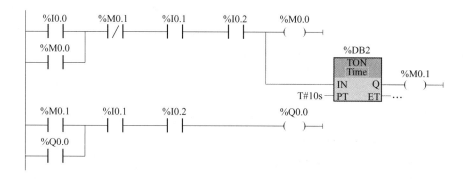

Figure 3-11 Motor reference program for delayed start of PLC

1) The M0.1 normally closed contact is opened, which destroys the self-locking of the M0.0 coil. The M0.0 coil loses power, and the IN terminal of the timer is disconnected open and stop timing. The Q terminal of the timer is disconnected, and M0.1 is de-energized (the coil of M0.1 is only closed for one scanning cycle).

2) The normally open contact of M0.1 is closed, and the coil of Q0.0 is electrically locked, so the KA coil connected to the Q0.0 end of the PLC attracts.

The pull-in of the KA coil closes the KA normally-open contact in the main circuit, so that the KM coil of the contactor is powered on and pulled in. The attraction of the KM coil closes its main contacts and the motor starts.

(2) Stop immediately: Jogging SB_2, I0.1 normally open contacts first open and then close, so that the self-locking state of the Q0.0 coil is destroyed, and the Q0.0 coil is de-energized. The loss of power in the Q0.0 coil causes the intermediate relay KA coil to lose power and open. The power loss of the KA coil makes the normally open contact in the main circuit from closed to open, thereby de-energizing the relay KM coil, the main contact changes from closed to open, and the motor stops running.

When the main circuit overload causes the thermal relay FR to operate, the FR normally closes contact of the I0.2 terminal connected to the PLC is disconnected, and the motor stops.

Operation Steps

The operation steps are as follows:

(1) On the XK-SX2C advanced maintenance electrician training platform, use jumper wires to complete the wiring of the PLC control circuit, as shown in Figure 3-9, and the main circuit wiring shown in Figure 3-10.

(2) Use a multimeter to measure the main circuit and the control circuit separately, and check whether the circuit has short circuit, open circuit or wrong wiring. If there is a problem, carefully check until the fault is eliminated.

(3) Close the main power supply of the training platform, and turn on the computer and open the Portal software.

(4) Input the program, as shown in Figure 3-11 in the programming interface of Portal software, close the DC24V power switch powered by PLC, download the program to the PLC, set the PLC to the running state and monitor the program.

Task Presentation

The task presentation is as follows:

(1) Jogging SB_1, the intermediate relay KA coil is delayed to pull in for 10s.

(2) Jogging SB_2, the KA coil is immediately disconnected.

(3) Close the main circuit isolation switch QS.

(4) Jogging SB_1 again, the KA coil is delayed to pull in for 10s, and at the same time, the KM coil is pulled in and the motor starts.

(5) Jogging SB_2, the coils of KA and KM are immediately disconnected at the same time.

(6) After the demonstration, disconnect all power supplies.

Analysis and Summary

Summarize the task implementation process, find deficiencies, especially attach importance to the occurrence of malfunctions, and record in Table 3-3 after analysis and summary.

3.1.5 Task Evaluation

Score the completion of the wiring and debugging tasks of the motor's delayed start main control circuit, and fill the scoring results in Table 3-2.

Table 3-2 Score table

Task content	Assessment requirements	Grading	Distribution	Deduction	Score
Relay wiring and analysis	Correctly complete the wiring, demonstration and description of the main circuit and control circuit of the relay	(1) 10 points will be deducted for delayed start cannot be achieved; (2) Unable to stop deducting 5 points immediately; (3) 5 points will be deducted for each place where the operation phenomenon cannot be analyzed and explained	30		
PLC main circuit and control circuit wiring	Correctly complete the wiring of the main circuit and PLC control circuit	Complete the wiring of the main circuit and the PLC control circuit correctly, and 10 points will be deducted for each wrong connection	25		

Continued Table 3-2

Task content	Assessment requirements	Grading	Distribution	Deduction	Score
PLC programming and download	Input, download and monitor the PLC program according to the task requirements	(1) 15 points will be deducted when the software is unskilled and cannot complete the input operation of the program deducted 15 points; (2) 5 points will not be deducted for the IP address setting and downloading of the program; (3) 5 points will not be deducted for the stop and running status of the PLC controlled by Portal software; (4) 5 points will not be deducted for the operation of the Portal software monitoring program	15		
PLC control system operation demonstration	Demonstrate the start and stop of the motor correctly, and explain it with the program and hardware	(1) After jogging the start button, 10 points will be deducted when the student cannot correctly explain the motor's delayed start process; (2) After jogging the stop button, 10 points will be deducted when the student cannot correctly explain that the motor's immediate stop process	20		
Safe and civilized production	Abide by the 8S management system and the safety management system.	(1) 10 points will be deducted for not wearing labor protection articles; (2) There are hidden safety hazards in the operation, 5 points will be deducted each time for hidden safety hazards in the operation, until the deduction is completed; (3) The operation site was not sorted out and rectified in time. 2 points will be deducted each time until the deduction is completed	10		
		Total score			

3.1.6 Task Summary

The task implementation process record sheet is shown in Table 3-3.

Table 3-3 Task implementation process record sheet

	Task implementation process record sheet	
Fault 1	Fault phenomenon	
	Cause of failure	
	Troubleshooting process	
Fault 2	Fault phenomenon	
	Cause of failure	
	Troubleshooting process	
Fault 3	Fault phenomenon	
	Cause of failure	
	Troubleshooting process	
	Summary	

3.1.7 Task Development

Task requirements: Jogging the start button SB_1, the motor starts immediately, and stops after running for 8s. If the button SB_2 is jogged while the motor is running, the motor stops immediately. Please complete the following tasks in accordance with the above design requirements:

(1) Sign the main control circuit diagram of the relay control system and complete the wiring and debugging.

(2) Complete the PLC I/O address assignment.

(3) Design the main control circuit diagram of the PLC control system, and complete the wiring.

(4) Write PLC control program, complete program download, operation and software and hardware debugging.

Task 3.2 Design and Implementation of Main Circuit and Control Circuit of Motor's Delayed Stop

3.2.1 Task Description

This task uses the relay system and the PLC control system to complete the wiring and debugging of the three-phase asynchronous motor's delayed stop main control circuit.

3.2.2 Task Target

(1) Master the design, wiring and debugging of the main control circuit of the three-phase asynchronous motor delay stop relay control system.

(2) Understand the working principle of Siemens S7-1200 PLC set and reset instructions.

(3) Understand the working principle of Siemens S7-1200 PLC rising edge and falling edge

instructions.

(4) Master the delay of the three-phase asynchronous motor to stop the wiring of the main control circuit of the PLC control system.

(5) Understand the design ideas of the PLC program and master the download and debugging of the program.

3.2.3 Task-related Knowledge

3.2.3.1 Set and Reset Instructions

Among the bit logic instructions, there are three sets of six instructions related to set and reset instructions, which are introduced below.

Set Output and Reset Output Instructions

The S (set output) and R (reset output) instructions set and reset the specified bit operand. If the S and R coils of the same operand are powered off at the same time, the signal state of the specified operand does not change. The main features of set output command and reset output command are memory and holding function.

As shown in Figure 3-12 (a), if the normally open contact of I0.4 is closed, Q0.5 becomes the 1 state and maintains this state. Even if the normally open contact of I0.4 is opened, Q0.5 still maintains the 1 state. The normally open contact of I0.5 is closed, and Q0.5 becomes 0 state and maintains this state. Even if the normally open contact of I0.5 is opened, Q0.5 still maintains the 0 state. Figure 3-12 (b) is the corresponding timing chart of the program.

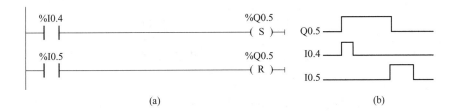

Figure 3-12 S/R instruction application example (1)

The program, as shown in Figure 3-13, achieves the same control effect as the program shown in Figure 3-11. I0.0 sets M0.0 to 1 to make the normally open contact of M0.0 close, and the IN terminal turns on to start timing. When the timing time reaches the timer setting value, the Q terminal of the timer is turned on, so that the M0.1 coil is energized, and the normally open contact of M0.1 is closed. The open contact of M0.1 closes, and resets M0.0. M0.0 normally open contact opens, and the timer stops counting, in addition, Q0.0 is set to 1 to start the motor. I0.1 and I0.2 use the normally closed form in the program because the two inputs in the PLC external circuit are connected to the normally closed form of the button and the thermal relay.

Pay attention to the following three points when using S and R instructions:

(1) Set and reset the same bit multiple times in one program.

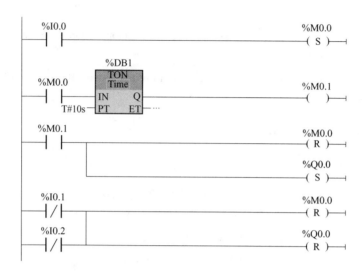

Figure 3-13 S/R instruction application example (2)

(2) When using S and R instructions, it is recommended not to connect the logic line for a long time, otherwise it will reduce the rigor of the program.

(3) It is not best to use coil output and set commands for the same coil in one program.

Set/Reset Trigger and Reset/Set Trigger

Figure 3-14 (a) shows the ladder diagram of the set/reset trigger, and the identifier is SR. Figure 3-14 (b) shows the ladder diagram of the reset/set flip-flop, and the identifier is RS.

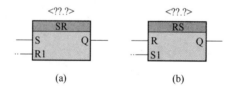

Figure 3-14 Ladder diagram form of SR and RS instructions

The SR instruction sets or resets the bit of the specified operand according to the signal states of the input S and R1. If the signal at input S is on and the signal state at input R1 is off, the specified operand is set to 1. If the signal input to S is off and the signal input to R1 is on, the specified operand is reset to 0. Input R1 has higher priority than input S. When both the signal states of inputs S and R1 are on, the signal state of the specified operand is reset (reset priority). If the signal states of both inputs S and R1 are off, the instruction will not be executed. Therefore, the signal state of the operand remains unchanged. The program example is shown in Figure 3-15. If the I0.0 and I0.1 input terminals are connected to the normally open buttons SB_1 and SB_2 respectively, then jogging SB_1, M0.1 and Q0.0 are set to 1; Jogging SB_2, M0.1 and Q0.0 are reset; If SB_1 and SB_2 are pressed at the same time, M0.1 and Q0.0 are reset. Both M0.1 and Q0.0 are operands of the SR instruction, in which coil Q0.0 is optional.

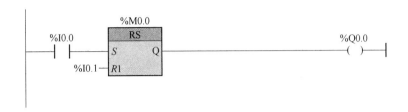

Figure 3-15 SR instruction application example

The RS instruction resets or sets the bit of the specified operand according to the signal states of the input R and S1. If the signal state of input R is on and the signal state of input S1 is off, the specified operand will be reset. If the signal state of input R is off and the signal of input S1 is on, the specified operand is set to 1. Input S1 has higher priority than input R. When the signal states of inputs R and S1 are both on, the signal state of the specified operand is set to 1 (set priority). If the signal states of both inputs R and S1 are off, the instruction will not be executed, so the signal state of the operand remains unchanged. The program example is shown in Figure 3-16. If the input terminals of I0.0 and I0.1 are connected to the normally open buttons SB_1 and SB_2, respectively, SB_1, M0.0 and Q0.0 will be reset when inching; SB_2, M0.0 and Q0.0 will all be set when inching; Press SB_1 and SB_2 at the same time, then M0.0 and Q0.0 are both set. Both M0.0 and Q0.0 are the operands of the RS instruction, and the coil Q0.0 is optional.

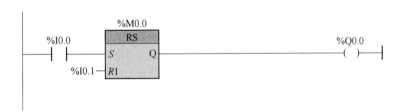

Figure 3-16 RS command application example

The design requirements of an application example of the SR instruction are as follows:

(1) If the responder has three input buttons, connect I0.0, I0.1 and I0.2 respectively.

(2) The output has three indicator lights, connected to Q4.0, Q4.1 and Q4.2, and the reset input button is connected to I0.4.

Requirement: Any one of the three people can answer the question, whoever presses the button first, whose indicator light is on first, and only one light can be lit. When the next question is asked, the host presses the reset button and the answer starts again.

The PLC program is shown in Figure 3-17. Assuming that the button connected to I0.0 is jogged first, Q0.4 is energized, the corresponding indicator light is on, and the normally closed contact of Q0.4 changes from closed disconnect. In this way, clicking the buttons corresponding to I0.1 and I0.2 will not execute the SR instruction which are connected in series.

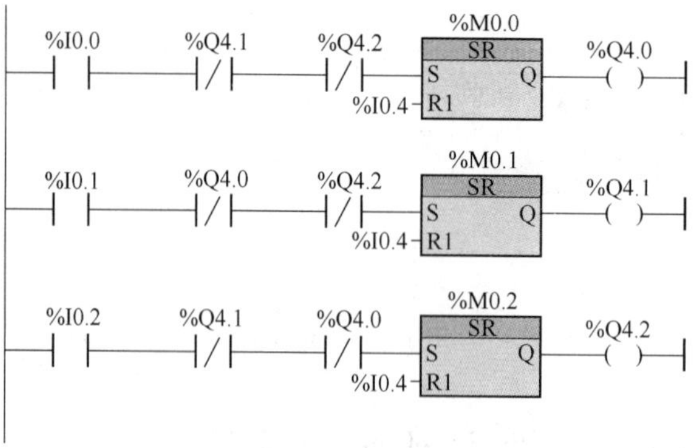

Figure 3-17 Reference program for answering device control PLC

Domain Instructions and Reset Area Instructions

The identifier of the set bit field instruction is SET_ BF, which can set multiple bits starting from a specific address. The identifier of the reset area instruction is RESET_ BF, which can reset multiple bits starting from a specific address. The program is shown in Figure 3-18. I0. 0 is turned on and off, so that 5 consecutive output coils starting from Q0. 0 are set to 1 (Q0. 0~Q0. 4 are set to 1). The on-off of I0. 1 resets the two output coils starting from Q0. 2, and the final running result is that only Q0. 0, Q0. 1, and Q0. 4 are set to 1.

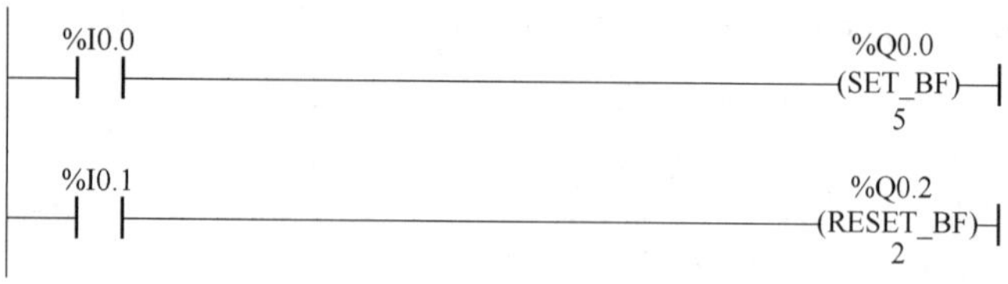

Figure 3-18 Application example of SET_ BF and RESET_ BF

Practice

Use the SR and RS instructions to complete the program design, as shown in Figure 3-11.

3. 2. 3. 2 Rising Edge Instruction and Falling Edge Instruction

A transition edge is generated when the signal state changes. When it changes from 0 to 1, a rising edge (or positive transition) is generated; If it changes from 1 to 0, a falling edge (or negative transition) is generated. As shown in Figure 3-19, when the normally open button SB connected to I0. 0 of the PLC is jogged, the I0. 0 state changes from 0 to 1 will produce a rising edge, producing a falling edge from 1 to 0.

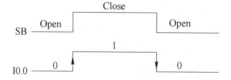

Figure 3-19 Schematic diagram of transition edge

In the S7-1200 PLC instruction system, there are four groups of eight instructions related to edge detection. Only two groups of instructions are introduced here.

Operand Signal Scanning Instruction

The operand signal scan instruction includes a scan operand signal rising edge instruction and a scan operand signal falling edge instruction. This group of instructions is used to detect the rising edge signal and falling edge signal of the operand, and at the same time, make the logic line turn on for one scan cycle, as shown in Figure 3-20.

```
    %I0.0                                    %Q0.0
   ─┤P├────────────────────────────────────(SET_BF)─
    %M4.3                                      4
    %I0.1                                    %Q0.2
   ─┤N├────────────────────────────────────(RESET_BF)─
    %M4.4                                      2
```

Figure 3-20   Application example of operand signal scanning instruction (1)

The instruction with P in the middle is the rising edge instruction of the scan operand signal. On the rising edge of I0.0, the contact is turned on for one scan cycle and Q0.0~Q0.3 is set to 1, and M4.3 is the edge storage bit used to store the state of I0.0 during the last scan cycle. By comparing the status of the two cycles before and after I0.0, the edge of the signal is detected. The address of the edge storage bit can only be used once in the program. Temporary local datas or I/O variables of code blocks aren't be used as edge storage bits. The instruction with N in the middle is the signal falling edge instruction of the scan operand. On the falling edge of I0.1, the coil of RESET_ BF is 'Energized' for one scan cycle, resetting Q0.2 and Q0.3, and M4.4 below the contact is the edge storage bit.

The program shown in Figure 3-21 realizes that when the normally open button connected to I0.0 is closed, Q0.0 is immediately energized, and when the normally open button connected to I0.0 is disconnected, Q0.0 is delayed by 5s Loss of power.

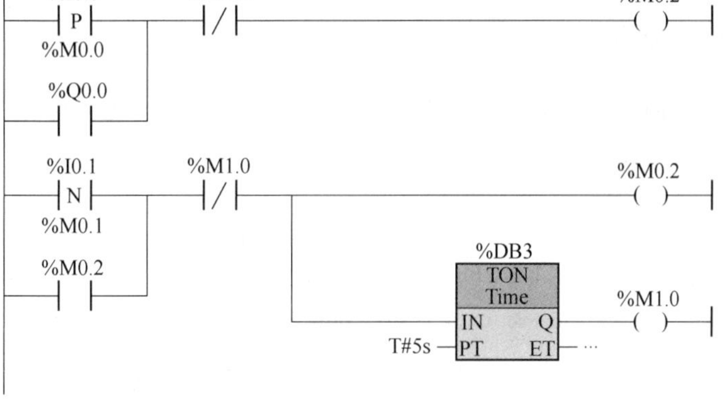

```
    %I0.0        %M1.0                              %M0.2
   ─┤P├─────────┤/├──────────────────────────────────( )─
    %M0.0
    %Q0.0
   ─┤ ├─

    %I0.1        %M1.0                              %M0.2
   ─┤N├─────────┤/├──────────────────────────────────( )─
    %M0.1
                                    %DB3
    %M0.2                           TON
   ─┤ ├─                            Time            %M1.0
                                 ──IN    Q──────────( )─
                            T#5s ──PT   ET──…
```

Figure 3-21   Application example of operand signal scanning instruction (2)

Scanning RLO Signal Edge Instructions

In Siemens PLC, bit 1 of the status word is called RLO (Result of Logic Operation). This bit stores the result of the logic instruction or arithmetic comparison instruction. In the logical row, the state of the RLO bit can represent information about the signal flow. The RLO state being 1, indicates that there is signal flow (on); When being 0, it is indicated that there is no signal flow (off). RLO can be used to trigger jump instructions.

In S7-1200 PLC, the instructions to scan the RLO signal edge are the RLO signal rising edge instruction (P_TRIG) and the RLO signal falling edge instruction (N_TRIG). As shown in Figure 3-22, when the power flow (RLO) flowing into the CLK input terminal of the P_TRIG instruction changes from off to on (rising edge), the output pulse width of the Q terminal of the P_TRIG instruction is one scan period power flow (turning on a scan cycle). Setting Q0.0 to 1, M8.0 below the instruction is the pulse storage bit. When the energy flow flowing into the CLK input terminal of the N_TRIG instruction is on to off (falling edge), the Q terminal of the N_TRIG instruction outputs a scanning cycle of energy flow (turning on a scanning cycle). Resetting Q0.0 to 0, and M8.1 below the instruction is a pulse memory bit. The P_TRIG instruction and N_TRIG instruction cannot be placed at the beginning and end of a logical line.

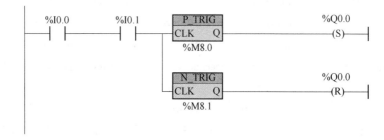

Figure 3-22 P_TRIG instruction and N_TRIG instruction application examples

Practice

Use the P_TRIG instruction and N_TRIG instruction to achieve the same function as the program in Figure 3-21.

3.2.4 Task Implementation

3.2.4.1 Wiring and Debugging of Main Control Circuit of Delayed Stop Control of Motor (Relay Control System)

Task Requirements

Jogging the start button, start the motor immediately; Jogging the stop button, stop the motor with a delay of 10s; Jogging the emergency stop button, stop the motor immediately. Completing the wiring and debugging of the main circuit and the control circuit, the circuit should have the necessary short-circuit and overload protection measures.

Finally, score the project after completion, as shown in Table 3-5.

Task Analysis

Pressing the start button, the contactor coil that controls the operation of the motor should immediately pull in, so that its main contact is closed to start the motor. To achieve the delay of the motor by pressing the stop button, this operation must be used to start timing after the timer is turned on. In order to ensure that the time relay can be energized to 10s, it needs to be achieved by the self-locking of the intermediate relay. When the timing time expires, the normally closed delay contact of the time relay is opened, the contactor coil controlling the operation of the motor is de-energized, and the motor is stopped. The main circuit should have an emergency stop button, a fuse and a thermal relay, and the control circuit should have a fuse and a normally closed contact of the thermal relay.

In summary, the use of time relays to achieve the motor delay start main control circuit is shown in Figure 3-23.

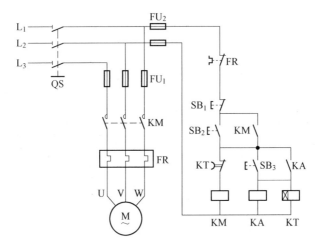

Figure 3-23   Main control circuit diagram of motor delay stop relay

Operation Steps

The operation steps are as follows

(1) On the XK-SX2C advanced maintenance electrician training platform, use jumper wires to complete the connection between the main circuit and the control circuit shown in Figure 3-23.

(2) Set the action time of the time relay to 10s.

(3) Use a multimeter to measure the main circuit and the control circuit separately to check whether the circuit has short circuit, open circuit or wrong connection. If there is a problem, check carefully until the fault is eliminated.

Task Presentation

The task presentation is as follows:

(1) Close the main power supply of the training platform and the circuit isolation switch QS.

(2) Jogging SB_2, the motor starts immediately.

(3) Jogging SB_3, the intermediate relay coil pulls in, the time relay starts timing, its corresponding indicator light is turned on, and the contactor coil is powered off and released after 10s.

Then the motor stops, and the intermediate relay and time relay are powered off.

(4) Jogging SB_2 again, the motor starts immediately. Jogging SB_1, the motor stops immediately.

(5) Disconnect the isolation switch QS and the main power supply of the training equipment.

3.2.4.2 Delayed Stop Motor Control Wiring and Debugging of Main Circuit and Control Circuit (PLC Control System)

Task Requirements

Jogging the start button connected to the PLC input point, the motor starts immediately; Jogging the stop button connected to the PLC input point, the motor delays for 10s; When the motor is running, Jogging the emergency stop button, the motor stops immediately. Completing the wiring and debugging of the main circuit and PLC control circuit, the circuit should have the necessary short-circuit and overload protection measures.

Finally, score the project after completion, as shown in Table 3-5.

Task Analysis

When designing a PLC control system, you should firstly analyze according to the design requirements and determine the PLC I/O address allocation, as shown in Table 3-4.

Table 3-4 I/O address allocation table of PLC control system with delayed motor stop

| \multicolumn{5}{c|}{Input signal} | \multicolumn{5}{c}{Output signal} |
No.	Function	Element	Address	No.	Controlled object	Element	Address
1	Start button	SB_1	I0.0	1	Intermediate relay	KA	Q0.0
2	Stop button	SB_2	I0.1	—	—	—	—
3	Emergency button	SB_3	I0.2	—	—	—	—
4	Overload protection	FR	I0.3	—	—	—	—

On the basis of completing the PLC address allocation table, the PLC external wiring diagram is design as shown in Figure 3-24.

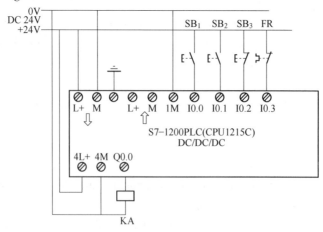

Figure 3-24 Wiring diagram of PLC control circuit for motor delay stop

In the main circuit, the control of the contactor coil needs to be achieved through the normally open contact of the intermediate relay, as shown in Figure 3-25.

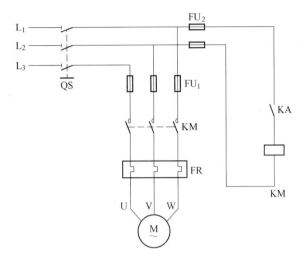

Figure 3-25   Wiring diagram of main circuit of motor delayed stop (PLC)

Comparing Figure 3-25 and Figure 3-10, It can be seen that these two circuit diagrams are exactly the same, which is also one of the advantages of the PLC control system relative to the relay control system: the PLC external wiring diagram and the main circuit are not modified or partially modified, and then the control function can be changed.

The PLC program is designed on the basis of the completion of the design of the PLC main circuit and control circuit. This project gives three design methods, as shown in Figures 3-26~3-28 respectively.

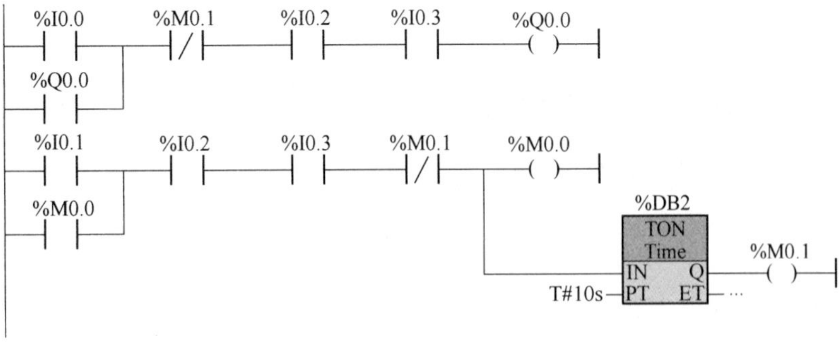

Figure 3-26   PLC reference program 1

The program is useded, as shown in Figure 3-27, to analyze the PLC program execution and hardware device operations, as follows:

(1) Start immediately: Jogging SB_1, Q0.0 is set to 1, and KA coil pulls in. The pull-in control of the KA coil controls the closing of the KA normally open contact in the main circuit, so that

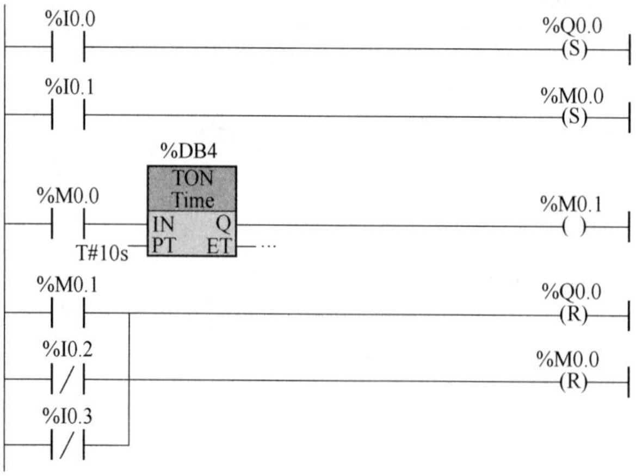

Figure 3-27 PLC reference program 2

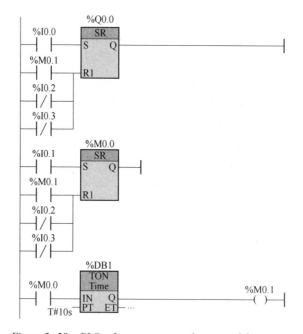

Figure 3-28 PLC reference program for motor delay stop 3

the KM coil of the contactor is powered on and pulled in. The attraction of the KM coil closes its main contacts and the motor starts.

(2) Delayed stop: Jogging SB_2, I0.1 is turned on and then off, the M0.0 coil is set to 1, the M0.0 coil is set to make its normally open contact closed, and the timer starts counting. When the timer counts to 10s, the Q terminal of the timer is turned on, so that the M0.1 coil is turned on. Since the M0.1 coil is turned on, Q0.0 is reset, and the KA coil is powered off and released. The release of the KA coil disconnects the KA normally open contact in the main circuit, and the KM coil of the contactor is powered on and off. The release of the KM coil disconnects its

main contact and the motor stops.

(3) Stop immediately: Jogging SB_3, I0.2 normally closed contact is closed after it is turned on. Resetting Q0.0 and M0.0, the motor stops running. When the main circuit overload causes the thermal relay FR to operate, the FR normally closed contact of the I0.3 terminal of the PLC is disconnected, and the motor stops.

Operation Steps

The operation steps are as follows:

(1) On XK-SX2C advanced maintenance electrician training equipment, use jumper wires to complete the wiring of the PLC control circuit shown in Figure 3-24 and the main circuit wiring shown in Figure 3-25.

(2) Use a multimeter to measure the main circuit and the control circuit separately, and check whether the circuit has short circuit, open circuit or wrong wiring. If there is a problem, carefully check until the fault is eliminated.

(3) Close the main power supply of the training equipment, turn on the computer and open the Portal software.

(4) Input the program shown in Figure 3-26 in the programming interface of Portal software, close the DC24V power switch powered by PLC, download the program to the PLC, set the PLC to the operating state and monitor the program.

(5) By jogging SB_1, SB_2 and SB_3, observe the state of the coil, the operation of the motor and the operation of the program.

(6) Download and monitor the PLC programs shown in Figures 3-27 and 3-28 respectively, and observe the running process of the program through the operation described in the steps of (5).

Task Presentation

The task presentation is as follows:

(1) Jogging SB_1, the intermediate relay coil KA will pull in immediately.

(2) Jogging SB_2, the KA coil is powered off after a delay of 10s.

(3) Jogging SB_1 again, the KA coil will pull in immediately.

(4) Jogging SB_3, KA coil is immediately powered off and released.

(5) Close the main circuit isolation switch QS.

(6) Jogging SB_1, the intermediate relay coil KA and the contactor coil KM immediately pull in, and the motor starts immediately.

(7) Jogging SB_2, KA coil and KM coil delay 10s power off and release, and the motor stops.

(8) Jogging SB_1 again, the motor starts immediately.

(9) Jogging SB_3, the motor stops immediately.

(10) After the demonstration, disconnect all power supplies.

Analysis and Summary

Summarize the entire task implementation process, find deficiencies, especially attach importance to the occurrence of faults, and record in Table 3-6 after analysis and summary.

3.2.5 Task Evaluation

Score the wiring of the main circuit and the control circuit of the motor delay stop control and the completion of the debugging task, and fill the scoring results in Table 3-5.

Table 3-5 Score table

Task content	Assessment requirements	Grading	Distribution	Deduction	Score
Wiring and demonstration of relay control circuit	Correctly complete the wiring, demonstration and phenomenon analysis of the main circuit and control circuit of the relay	(1) 5 points will be deducted when the circuit cannot be started immediately; (2) 10 points will be deducted when the circuit cannot realize the delay stop; (3) 5 points will be deducted when the circuit cannot realize the emergency stop function; (4) 5 points will be deducted for each place where the operation phenomenon cannot be analyzed and explained	30		
Main control circuit wiring of PLC control system	Correctly complete the wiring of the main circuit and PLC control circuit	10 points will be deducted when the wiring of the main circuit and PLC control circuit is deducted for each wrong connection	20		
Writing and downloading of PLC program	Input, download and monitor the PLC program according to the task requirements	(1) 15 points will be deducted when the software is unskilled and cannot complete the input operation of the program; (2) 5 points will not be deducted for the IP address setting and downloading of the program; (3) 5 points will not be deducted for the stop and running status of the PLC controlled by Portal software; (4) 5 points will not be deducted for the operation of the Portal software monitoring program	20		
Operation demonstration of PLC control system	Demonstrate the start and stop of the motor correctly, and explain it with the program and hardware	(1) 5 points will be deducted for not correctly explain the motor's immediate start process after jogging the start button; (2) 10 points will be deducted for not correctly explaining the motor's delay stop process after jogging the stop button; (3) 5 points will be deducted for not correctly explaining the motor's immediate stop process when jogging the emergency stop button	20		

Continued Table 3-5

Task content	Assessment requirements	Grading	Distribution	Deduction	Score
Safe and civilized production	Abide by the 8S management system and the safety management system	(1) 10 points will be deducted for not wearing labor protection articles; (2) 5 points will be deducted each time when hidden safety hazards in the operation, until the deduction completed; (3) 2 points will be deducted each time when the operation site is not sorted out and rectified in time, until the deduction completed	10		
		Total score			

3.2.6 Task Summary

The task implementation process record sheet is shown in Table 3-6.

Table 3-6 Task implementation process record sheet

	Task implementation process record sheet	
Fault 1	Fault phenomenon	
	Cause of failure	
	Troubleshooting process	
Fault 2	Fault phenomenon	
	Cause of failure	
	Troubleshooting process	
Fault 3	Fault phenomenon	
	Cause of failure	
	Troubleshooting process	
	Summary	

3.2.7 Task Development

Task requirements: Jogging the start button SB_1, the motor will start immediately after a delay of 5s; Jogging the delay stop button SB_2, the motor will stop after a delay of 8s. If the emergency stop button SB_3 is jogged while the motor is running, the motor stops immediately. Please complete the following tasks in accordance with the above design requirements:

(1) Design the main control circuit diagram of the relay control system and complete the wiring and debugging.

(2) Determine the PLC I/O address assignment.

(3) Design the main control circuit diagram of the PLC control system, and complete the wiring.

(4) Use two methods to write the PLC control program to complete the program download, operation and software and hardware debugging.

Task 3.3 Design and Implementation of Main Circuit and Control Circuit of Cyclic Start and Stop Control of Motor

3.3.1 Task Description

The purpose of this task is to use the relay system and PLC control system to realize the wiring and debugging of the main circuit and control circuit of the three-phase asynchronous motor cycle start and stop control.

3.3.2 Task Target

(1) Master the wiring and debugging of the main control circuit of the relay control system of the cyclic start and stop control of the three-phase asynchronous motor.

(2) Understand the working principle of Siemens S7-1200 PLC comparison instruction.

(3) Understand the working principle of Siemens S7-1200 PLC counter instructions.

(4) Master the wiring of the main control circuit of the PLC control system of the cyclic start and stop control of the three-phase asynchronous motor.

(5) Understand the design ideas of the PLC program and master the download and debugging of the program.

3.3.3 Task-related Knowledge

3.3.3.1 Comparison Instructions

In S7-1200 PLC, the instructions related to the comparison have two operands of equal comparison, not equal comparison, greater than comparison, less than comparison, range comparison, variable comparison, etc. Here only six commonly used comparison instructions are analyzed, as shown in Table 3-7. The comparison instruction can realize the comparison of integers, double integers, real numbers, character strings, etc. When the comparison result is true (established), the comparison instruction is turned on.

Table 3-7 Comparison instruction ladder diagram form and connection condition

Instruction	Logical line connect condition	Instruction	Logical line connect condition
─┤ IN1 == ??? IN2 ├─	IN1 = IN2	─┤ IN1 >= ??? IN2 ├─	IN1 ⩾ IN2
─┤ IN1 <> ??? IN2 ├─	IN1 ≠ IN2	─┤ IN1 <= ??? IN2 ├─	IN1 < IN2

Continued Table 3-7

Instruction	Logical line connect condition	Instruction	Logical line connect condition
⊣ IN1 > ??? IN2 ⊢	IN1>IN2	⊣ IN1 <= ??? IN2 ⊢	IN1≤IN2

In PLC basic programming, comparison instructions related to timers and counters are more commonly used, and they can be the comparison of the current values of two timers (or counters). It can also be a comparison between the current value of a timer (or counter) and a constant. Appropriate use of comparison instruction programming can simplify the program. As shown in Figure 3-29, if the I0.0 connection button is normally open SB_1, the I0.1 connection button is normally open SB_2. Jogging SB_1, Q0.0 turns on after 1s delay, Q0.1 turns on after 3s delay, and Q0.2 turns on after 5s delay; Jogging SB_2, Q0.2 turns off immediately, Q0.1 delays 1s to disconnect, and Q0.0 delays 2s to disconnect. In the program, the 5s timer is named T1, the current value of the T1 timer is 'T1.ET', and the current value of the 2s timer is 'MD10'. Both methods can read the current value of the timer value.

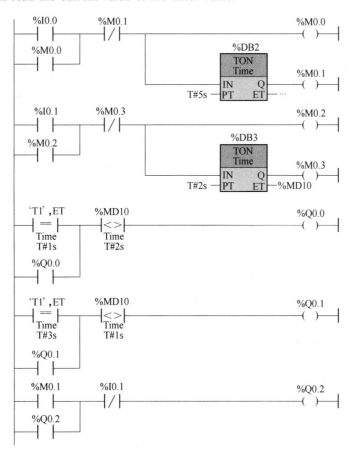

Figure 3-29 Comparison instruction example (1)

Sometimes one or several output points need to be turned on and off cyclically. If you use the compare command in programming, you can also simplify the program writing in the program shown in Figure 3-30. I0. 1 connection button normally open SB_2; Q0. 0 and Q0. 1 are connected to an indicator light HL_1 and HL_2. When SB_1 is jogged, HL_1 turns on immediately, turning off after 5s. HL_1 turns off for 2s, HL_2 turns on. After 4s, then HL_2 turns off, and after 1s, HL_1 turns on again, and so on. When SB_2 is jogged, HL_1 and HL_2 immediately go out. In the program, the logic line before the timer T1 (12-second timer) is connected to the M0. 1 normally closed contact in order to realize the current value of the timer to cycle between 0~12s.

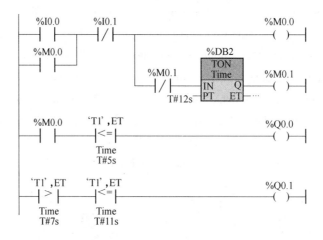

Figure 3-30 Comparison command example (2)

Practices

(1) Please program the program shown in Figure 3-29 with set and reset instructions.

(2) Please program the program shown in Figure 3-30 with the instruction equal to and not equal to the instruction.

3. 3. 3. 2 Counter Instructions

S7-1200 has three types of counters: up counter (CTU), down counter (CTD) and up/down counter (CTUD). They are software counters, and their maximum count rate is limited by the execution rate of the OB in which they are located. If you need a higher rate counter, you can use the high-speed counter built into the CPU. When calling the counter instruction, you need to generate an instance data block to store counter data.

Up Counter Instruction

The identifier of the up counter is CTU. The instruction has five parameters:

(1) Counting input CU: When the value at the CU terminal changes from 0 to 1 (signal rising edge), the current count value of the CTU is increased by 1, up to 32767.

(2) Reset input R: When the value of R terminal changes from 0 to 1 (signal rising edge), the current CTU count value returns to 0.

(3) Set value input PV: Counter set value input terminal.

(4) Counter status output Q: The signal status of output Q is determined by the parameter PV. If the current counter value is greater than or equal to the value of parameter PV, the signal state of output Q is set to 1 (ON).

(5) Current counter value CV: The current count value of the counter is stored.

The working principle of the up-counter instruction is shown in Figure 3-31, in which Figure 3-31 (a) is the PLC program, and Figure 3-31 (b) is the program execution timing diagram.

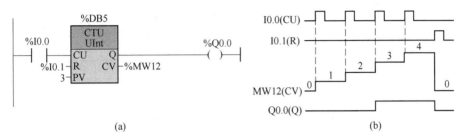

Figure 3-31 Working principle of count up instruction

Down Counter Instruction

The identifier of the down counter is CTD. The instruction has five parameters:

(1) Counting input CD: When the value at the CU terminal changes from 0 to 1 (signal rising edge), the current count value of the CTD is decremented by 1 to a minimum of 32768.

(2) Load input LD: When the value at the LD terminal changes from 0 to 1 (signal rising edge), CTD loads the count value set value PV, and the current value of the counter is the value set by PV.

(3) Set value input PV: Set value input terminal is countered.

(4) Current counter value CV: The current count value of the counter is stored.

(5) Counter status output Q: When the value of parameter CV is equal to or less than 0, the counter sets the signal status of output Q to 1 (ON).

The working principle of the down-counter instruction is shown in Figure 3-32, where Figure 3-32 (a) is the PLC program, and Figure 3-32 (b) is the program execution timing diagram.

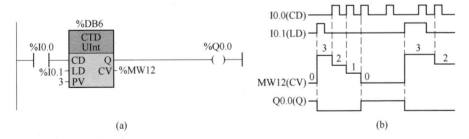

Figure 3-32 Working principle of count down instruction

Up-down Counter Instruction

The identifier of the up-down counter is CTUD. The instruction has functions of up-counting and down-counting. The instruction has eight parameters. Among them, QU is the count-up status output and QD is the count-down status output. For other parameter descriptions, the parameter descriptions of the up counter and down counter are refered to. The working principle of the up-down counter instruction is shown in Figure 3-33, where Figure 3-33 (a) is the PLC program, and Figure 3-33 (b) is the program execution timing diagram.

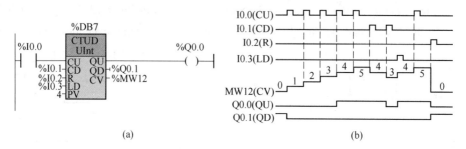

Figure 3-33 Working principle of count up and down instruction

The application of counter instructions is more common, such as the counting of the number of products on the production line, the counting of the number of motor runs and the number of stop cycles, etc. The PLC program, as shown in Figure 3-34, realizes the turning on and off of Q0.0 with one button. Among them, the Q terminal of the CTU is connected to the M0.0 coil. Its function is to make the counter count to the set value 2. After the M0.0 coil is energized, the normally open contact is closed, and the counter is reset to achieve operation. "IEC_Counter_0_DB". CV is the current value of the counter.

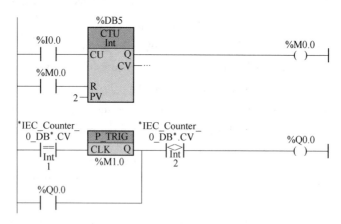

Figure 3-34 Counter example (1)

Figure 3-35 is another programming method to achieve this function. In this program, if the current value of the CTD is 0, at the beginning of PLC operation, the M0.0 coil will be energized due to the connection of the Q terminal of the CTD, and its normally open contact will be closed. The value 3 is loaded and the current value of the counter is 3. When the button corresponding to I0.0 is jogged twice, the current value of the counter is 1, the comparison command on the LD terminal is turned on, and the set value of the CTD is loaded again.

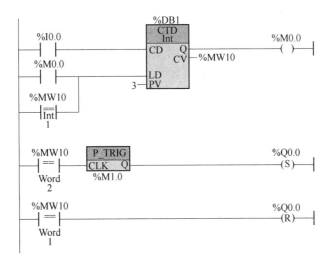

Figure 3-35 Counter example (2)

For relatively complex design requirements, multiple counters may need to be used in programming. The structure of the ballot box is shown in Figure 3-36. There are four buttons $SB_1 \sim SB_4$ (corresponding to I0.0~I0.3 of PLC) on the ballot box, which correspond to A, B, abstention and reset; The indicator lights $HL_1 \sim HL_3$ (corresponding to Q0.0~Q0.2 of PLC) correspond to the indications that A wins, B wins and equal votes. Before each vote, you must click the reset button to reset the counter to zero. There are ten students in a class. A and B are running for class monitors. A and B also participate in the voting. Only after the voting is finished, the voting results are displayed. Whoever has the most votes is elected, and the corresponding indicator lights.

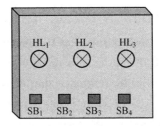

Figure 3-36 Schematic diagram of the ballot box

Here, two methods are used for programming and design in the program shown in Figure 3-37. In this program, three CTUs are used to achieve the design requirements. Among them, DB2 corresponds to CTU calculation A's votes; DB3 corresponds to CTU calculation the number of votes obtained by B; DB4 corresponds to the CTU to calculate the total number of votes. When the voting is over, at the end of the block, the value of MW14 is equal to 10, and the comparison command is turned on, so that the comparison of the number of votes received by A and B is reflected in the state of the corresponding indicator. The PV values of the three counters can be set arbitrarily within the setting range, because only their current values are read in the program, and the Q function is not used.

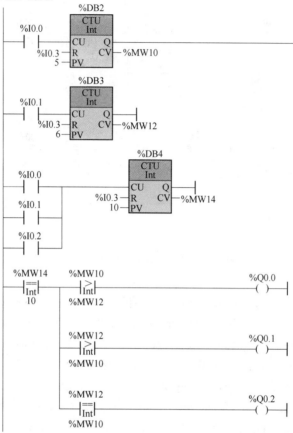

Figure 3-37 PLC program design of the ballot box (1)

You can also use a CTU and a CTUD to achieve the design requirements, as shown in Figure 3-38. In this design method, CTU counts the total number of votes, and CTUD counts up or down the votes obtained by A and B. It can be seen that, from the program, since the counter is reset by jogging the reset button SB_4, the current value of CTUD becomes 0. After the voting starts, If A gets 1 vote, CTUD increases by 1; If B gets 1 vote, CTUD decrements by 1. When the vote ends, we can know who has the most votes by comparing the current value of the counter with 0.

Practice

Design the PLC program of the ballot box with CTD and CTUD.

3.3.4 Task Implementation

3.3.4.1 Wiring and Debugging of Main Circuit and Control Circuit of Motor Cycle Start and Stop (Relay Control System)

Task Requirements

Jogging the start button, the motor will start immediately, running for 10s and stopping for 5s,

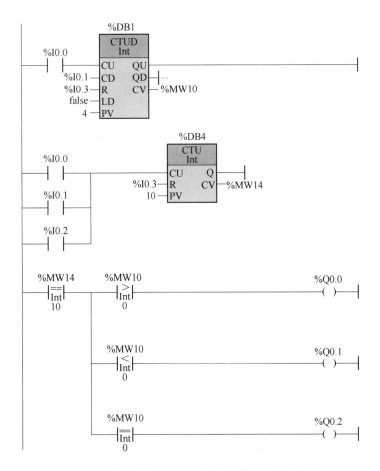

Figure 3-38　PLC program design of the ballot box 2

and then running in a cycle; Jogging the emergency stop button, the motor will stop immediately. Completing the wiring and debugging of the main circuit and the control circuit, the circuit should have the necessary short-circuit and overload protection measures.

Finally, score the project after completion, as shown in Table 3-9.

Task Analysis

Assume that the start button of the circuit is SB_1 and the stop button SB_2, the time relay corresponding to the motor running state is KT_1, and the time relay corresponding to the motor stop state is KT_2.

Pressing the start button SB_1, the control motor contactor coil KM. and KT_1 coil should be immediately engaged, so that the motor starts and starts timing. When the motor running time reaches 10s, the delay opening contact of KT_1 should make the KM coil and KT_1 coil de-energized and released. At the same time, the delayed closing contact of KT_1 should make the KT_2 coil pull in. To ensure that the KT_2 coil pull-in time can reach 5s, it needs to be achieved by the self-locking of the intermediate relay KA. At this time, its delayed breaking contact destroys its own self-locking state, and its delayed closing contact makes the contactor coil KM and KT_1

coil pull in again.

In summary, the use of time relays to achieve the motor delay start master control circuit is shown in Figure 3-39.

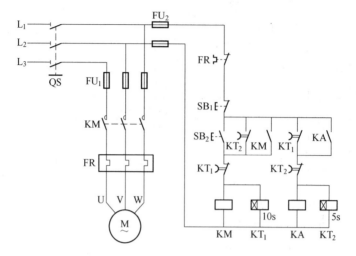

Figure 3-39 Main control circuit diagram of motor cycle start-stop relay

Operation Steps

The operation steps are as follows:

(1) On the XK-SX2C training device, use the jumper to complete the connection between the main circuit and the control circuit shown in Figure 3-39.

(2) Set the action time of KT_1 to 10s and the action time of KT_2 to 5s.

(3) Use a multimeter to measure the main circuit and the control circuit separately to check whether the circuit has short circuit, open circuit or wrong connection. If there is a problem, check carefully until the fault is eliminated.

Task Presentation

The task presentation is as follows:

(1) Close the total power supply of the training equipment and the circuit isolation switch QS.

(2) Jogging SB_2, the motor starts immediately.

(3) The motor stops after 10s of operation, restarting after 5s of stop, and then it runs cyclically.

(4) Jogging SB_1, the motor stops immediately.

(5) Disconnect the isolation switch QS and the main power supply of the training platform.

3.3.4.2 Wiring and Debugging of Main Control Circuit of Motor Cycle Start and Stop (PLC Control System)

Task Requirements

Only one button is used to start and stop the motor. When the button is jogged for the first time, the motor cycle runs as follows: The motor starts immediately, and it stops 5s after running for 10s, then it restarts and runs for 10s to stop for 5s; When the button is jogged twice, the motor stops immediately. Pressing the emergency stop button when the motor is running, the motor will stop immediately. Completing the wiring and debugging of the main circuit and PLC control circuit, the circuit should have the necessary short-circuit and overload protection measures.

Finally, score the project after completion, as shown in Table 3-9.

Task Analysis

When designing a PLC control system, you should analyze according to the design requirements firstly and determine the PLC I/O address allocation, as shown in Table 3-8.

Table 3-8 I/O address allocation table of PLC control system for motor cycle start and stop

	Input signal				Output signal		
No.	Function	Element	Address	No.	Controlled object	Element	Address
1	Start button	SB_1	I0.0	1	Intermediate relay	KA	Q0.0
2	Stop button	SB_2	I0.1	—	—	—	—
3	Overload protection	FR	I0.2	—	—	—	—

Based on the completion of the PLC address allocation table, the external wiring diagram of the design PLC is shown in Figure 3-40.

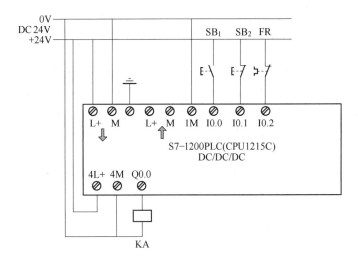

Figure 3-40 Wiring diagram of PLC control circuit for motor cycle start and stop

In the main circuit, the control of the contactor coil needs to be controlled by the normally open contact of the intermediate relay.

The PLC program is designed on the basis of the completion of the design of the PLC main circuit and the control circuit. This task gives two design methods. The first method does not use comparison instructions for programming. The reference program is shown in Figure 3-41. The M0.0 normally open contact in front of the counter in the program is to make the operation of the SB_1 button repeatable, that is, the motor can be started and stopped many times. Of course, the comparison instruction can also be used here to compare the current value of the counter with the set value 2. In block 2, the P_TRIG instruction is used to detect the rising edge of the counter when the current value of the counter is 1, a rising edge is generated, as long as the purpose is to turn on the Q0.0 self-locking condition one moment, not always on (If you don't jog SB$_1$ for the second time).

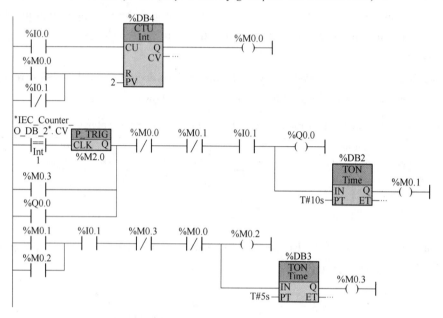

Figure 3-41 Motor cycle start and stop PLC reference program (1)

If you use the compare instruction to intercept the time period of the timer, the program will be simpler, as shown in Figure 3-42. This design circulates the current value of the timer between 0~15 s by adding an M0.2 normally closed contact before the timer, and makes the motor cycle start and stop by intercepting the first 10s with a comparison instruction.

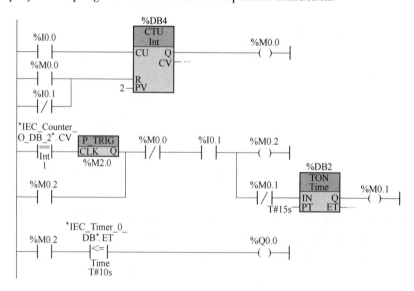

Figure 3-42 PLC reference program for motor cycle start and stop (2)

Operation Steps

The operation steps are as follows:

(1) On the XK-SX2C training device, use jumper wires to complete the wiring of the PLC control circuit shown in Figure 3-40, and the main circuit wiring shown in Figure 3-42.

(2) Use a multimeter to measure the main circuit and the control circuit separately, and check whether the circuit has short circuit, open circuit or wrong wiring. If there is a problem, check carefully until the fault is eliminated.

(3) Close the main power supply of the training platform, and turn on the computer and open the Portal software.

(4) Input the program shown in Figure 3-41 in the programming interface of the Portal software, close the DC 24V power switch powered by the PLC, download the program to the PLC, set the PLC to the operating state and monitor the program.

(5) By jogging SB_1 and SB_2, observe the state of the coil, the operation of the motor and the operation of the program.

(6) Download and monitor the PLC program shown in Figure 3-42, and observe the running process of the program through the operation described in the step of (5).

Task Presentations

The task presentations is as follows:

(1) SB_1 is jogged for the first time, and the intermediate relay coil KA is immediately engaged.

(2) The KA coil is powered off and released after a delay of 10s.

(3) After another 5s, the KA coil pulls in again.

(4) Jogging SB_1 again, KA immediately power off and release.

(5) In the state of KA coil pull-in, jogging SB_2, KA coil is immediately powered off and released.

(6) Close the main circuit isolation switch QS.

(7) For the first time to jog SB_1, the intermediate relay coil KA and the contactor coil KM pull in and the motor starts immediately.

(8) After 10s, the KA coil and KM coil are simultaneously powered off and released, and the motor stops.

(9) The motor starts again after 5s of stopping, and runs cyclically between the steps of (6) and (7).

(10) Jogging SB_1 again, the motor stops immediately.

(11) When the motor is running, jogging SB_2 and the motor will stop immediately.

(12) After the demonstration ends, disconnect all power supplies.

Analysis and Summary

Summarize the entire task implementation process, find deficiencies, especially attach importance to the occurrence of faults, and record in Table 3-10 after analysis and summary.

3.3.5 Task Evaluation

Score the wiring of the main circuit and the control circuit of the motor delay stop control and the completion of the debugging task, and fill the scoring results in Table 3-9.

Table 3-9 Score table

Task content	Assessment requirements	Grading	Distribution	Deduction	Score
Relay wiring and analysis	Correctly complete the wiring, demonstration and description of the main circuit and control circuit of the relay	(1) 5 points will be deducted when the motor cannot be started immediately; (2) The circuit cannot achieve 10s of operation and stop for 5s to deduct 10 points; (3) 5 points will be deducted when the motor cannot realize the emergency stop function; (4) 5 points will be deducted for each place where the operation phenomenon cannot be analyzed and explained	30		
Main circuit and control circuit wiring of PLC	Correctly complete the wiring of the main circuit and PLC control circuit	The wiring of the main circuit and PLC control circuit is deducted by 5 points for each wrong connection	15		
Programming and download of PLC	Input, download and monitor the PLC program according to the task requirements	(1) 15 points will be deducted when the software is unskilled and cannot complete the input operation of the program; (2) 5 points will not be deducted for IP address setting and download of the program; (3) 5 points will not be deducted for the stop and running status of the PLC controlled by the Portal software; (4) 5 points will not be deducted for the operation of the Portal software monitoring program	25		
Operation demonstration of PLC control system	Demonstrate the start and stop of the motor correctly, and explain it with the program and hardware	(1) Pressing the button SB_1 for the first time, it cannot correctly analyze the 5 points deduction of the motor's immediate start process; (2) Cannot combine software and hardware to analyze the working process of motor cycle start and stop 10 points deduction; (3) Jogging the emergency stop button, it cannot correctly analyze the 5 points deduction of the motor's immediate stop process	20		

Continued Table 3-9

Task content	Assessment requirements	Grading	Distribution	Deduction	Score
Safe and civilized production	Abide by the 8S management system and the safety management system	(1) 10 points will be deducted for not wearing labor protection articles; (2) 5 points will be deducted each time for hidden safety hazards in the operation, until the deduction completed; (3) 2 points will be deducted each time when the operation site was not sorted out and rectified in time, until the deduction completed	10		
		Total score			

3.3.6 Task Summary

The task implementation process record sheet is shown in Table 3-10.

Table 3-10 Task implementation process record sheet

	Task implementation process record sheet	
Fault 1	Fault phenomenon	
	Cause of failure	
	Troubleshooting process	
Fault 2	Fault phenomenon	
	Cause of failure	
	Troubleshooting process	
Fault 3	Fault phenomenon	
	Cause of failure	
	Troubleshooting process	
	Summary	

3.3.7 Task Development

3.3.7.1 Complete Design, Wiring and Debugging of Main Control Circuit of Two Motors Starting and Stopping in Sequence

Task requirements: For the first jogging SB_1, motor M_1 starts immediately, and motor M_2 starts after 5s; For the second jogging SB_1, M_2 stops immediately, and after 8s, motor M_1 stops. Jogging SB_2 at any time, M_1, M_2 will stop immediately.

According to the development requirements, the tasks are completed as follows:

(1) Design the main control circuit diagram of the relay control system, and complete the wiring and debugging.

(2) Determine the PLC I/O address allocation.

(3) Design the main control circuit diagram of the PLC control system, and complete the wiring.

(4) Use two methods to write the PLC control program to complete the program download, operation and software and hardware debugging.

3.3.7.2 Complete Design, Wiring and Debugging of PLC Main Control Circuit of a Motor Cycle Start and Stop

Task requirements: Jogging SB_1, the motor starts with a delay of 3s, stopping after 5s of operation, and then it stops the cycle of 2s. It stops automatically after 3 cycles. Jogging SB_2 at any time, the motor stops immediately.

According to the development requirements, the tasks are completed as follows:

(1) Determine the PLC I/O address allocation.

(2) Design the main control circuit diagram of the PLC control system, and complete the wiring.

(3) Write PLC control program, and complete program download, operation and debugging of software and hardware.

3.3.8 Project Summary

Through the study and practice of typical tasks in Project 3, the working principles and usage methods of S7-1200 PLC common instructions are understood, and the basic design methods of PLC programs are mastered. Through the comparative study of the relay control system and the PLC control system, it is not difficult to get the following conclusions:

(1) The logic of the relay control system uses hardware wiring, and the series or parallel connection of the mechanical contacts of the relay is used to form the control logic. The wiring is numerous and complicated, the volume is large, and the power consumption is large. After the system is completed, it is more difficult to change or add functions. In addition, the number of contacts of the relay is limited, so the flexibility and scalability of the electrical control system are greatly limited. The PLC uses computer technology, and its control logic is stored in the memory in the form of a program. If you need to change the control logic, we only need to change the program to easily change or increase systerm functions. The system has few connections, small size, low power consumption, and the soft relay of the PLC is essentially the state of the memory unit, so the number of contacts of the soft relay is unlimited, and the flexibility and scalability of the PLC system is good.

(2) The relay control system adopts the time delay action of the time relay to control the time, and the time delay precision of the time relay is not high. The PLC uses a semiconductor integrated

circuit as a timer, and the clock pulse is generated by a crystal oscillator, with high precision and a wide timing range. The user can set the timing value in the program as needed, and it is convenient to modify, and the PLC has a counting function, while the relay control system generally does not have a counting function.

(3) Because the electrical control system uses a large number of mechanical contacts, the reliability and maintainability of the entire system are poor. A large number of switching actions in the PLC control system are completed by non-contact semiconductor circuits, which have a long life and high reliability. The PLC also has a self-diagnosis function, which can detect its own faults, display it to the operator at any time, and it can dynamically monitor the execution of control procedures providing convenience for on-site debugging and maintenance.

Exercises

(1) Circuit Wiring Analysis Questions: Figure 3-43 shows the main circuit diagram of a relay for forward and reverse control of a certain motor.

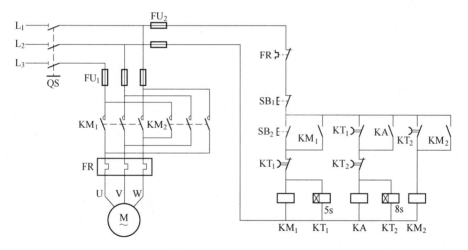

Figure 3-43 Motor control relay control circuit diagram

1) Analyze the control process of the circuit and fill in the corresponding component functions in Table 3-11.

Table 3-11 Circuit element function table

No.	Element	Function	No.	Element	Function
1	SB_1		4	KM_2	
2	SB_2		5	KA_1	
3	KM_1		6	KA_2	

2) When jogging SB_2 and SB_1, how will the motor work? and why?

3) Complete the wiring of the circuit, operate the corresponding button, and check whether

the analysis is correct.

(2) Motor positive and negative PLC main control circuit wiring and programming: On the basis of completing the main circuit design of the motor forward and reverse PLC control system and the PLC external circuit wiring diagram design, please write programs and download to the PLC for debugging according to the following design requirements, and download them to the PLC for debugging:

1) Jogging SB_1, the motor will start forward rotation after 2s delay, and stop after 3s of operation; Jogging SB_2, the motor will start reversely immediately, and stop after 5s. SB_3 is an emergency stop button.

2) Jogging SB_1, the motor rotates forward for 10s, stops for 5s, and then reverses. Jogging SB_2, the motor will stop after 2s delay. SB_3 is the emergency stop button.

3) Jogging SB_1, the motor rotates forward for 5s and stops for 2s, then reverses for 6s and stops for 3s, and then cycles in sequence. If jogging SB_2, the motor will stop for 3s with delay. SB_3 is the emergency stop button.

4) SB_1 is the mode selection button, SB_2 is the start button, and SB_3 is the emergency stop button. The mode selection must be made before starting: jogging SB_1 once, the motor operation mode is forward rotation and immediately start selection; Jogging SB_1 twice, motor operation mode is delayed 3s forward rotation start selection; Jogging SB_1 three times, the motor running mode is reverse to start immediately; Jogging SB_1 four times, the motor operation mode is reverse rotation and stops after 5s. After the operation mode is selected, jog SB_2 to start the motor. SB_3 is an emergency stop button. The number of jogs of SB_1 is cleared after Jogging SB_2.

(3) PLC programming and debugging of simulation control of the ventilation system of the production workshop: The ventilation system of an enterprise's production workshop is composed of fans of 1# and 2# and a status indicator, where 1# is the main fan, and 2# is the backup fan. Each fan has its own start button and stop button. The entire ventilation system has a start button and an emergency stop button. The control requirements are as follows:

1) Commissioning operation: Each fan can be started and stopped individually.

2) Normal operation: Pressing the system start button, the fan will start to run immediately, and the running indicator will go from off to steady on. Pressing the stop button in the running mode, the fan system will stop immediately, and the running indicator will go out.

3) Maintenance operation: Under normal operation state, jogging the 1# fan, the button is stop to make it switch from the working state to the stop operation state for maintenance. The operation indicator goes off, and 2# fan will automatically start running after 5s delay. After the 2# fan is running, the indicator light flashes at a frequency of 1Hz; After the 1# fan is maintained, click the stop button of the 2# fan, which stops immediately. The running indicator goes out, and the 1# fan is automatically started after a delay of 3s. The operation is put into normal operation, and the indicator light turns on.

Complete the following tasks according to the design requirements:

1) Complete the I/O address allocation of PLC according to the control requirements.
2) Design PLC external hardware wiring diagram.
3) Write PLC program and run debugging.

Project 4 Inverter Control of Three-phase Asynchronous Motor

Inverter is an adjustable speed drive device that uses variable frequency drive technology to change the frequency and amplitude of the working voltage of the AC motor to smoothly control the speed and torque of the AC motor to meet the wide demand of AC motor stepless speed regulation. Due to the excellent performance of the inverter in speed regulation and energy saving, it is widely used in various mechanical equipment control fields, such as machine tools, automated production lines, building elevators and so on. This project mainly learns how to use SINAMICS G120C inverter of German Siemens to control the operation of three-phase asynchronous motor. According to different control methods, the project includes the use of the G120C inverter operation panel to control the operation of the three-phase asynchronous motor and the use of the G120C PN inverter and S7-1200 PLC to control the operation of the three-phase asynchronous motor, gradually grasping the use of the inverter from shallow to deep.

Task 4.1 Using Inverter Panel to Control Operation of Three-phase Asynchronous Motor

4.1.1 Task Description

The basic operator panel BOP-2 of SINAMICS G120C inverter can be used for debugging, operation monitoring and input parameter setting of inverter. The menu navigating and related parameters displayed on the LCD screen can be used to debug the inverter. The local control of the inverter can be easily carried out through the navigation key, and the manual/automatic function switching can be completed through a special key.

This task uses the basic operator panel of BOP-2 of the SINAMICS G120C inverter to complete the settings of the inverter's common parameters, complete the control of the start and stop, forward and reverse operation, and speed regulation of the three-phase asynchronous motor.

4.1.2 Task Target

(1) Understand the basic structure and working principle of the inverter.

(2) Familiar with the functions of the operation panel and wiring terminals of Siemens G120C inverter.

(3) Master the simple installation and wiring method of Siemens G120C inverter.

(4) Master the setting method of common parameters of Siemens G120C inverter.

(5) Use the basic operator panel of BOP-2 of Siemens G120C inverter to control the start, stop, forward and reverse running, and speed control of three-phase asynchronous motor.

4.1.3 Task-related Knowledge

4.1.3.1 Overview of Inverter

Definition and Function of Inverter

Inverter is a power control device that uses frequency conversion technology and microelectronic technology to control the AC motor by changing the frequency of the motor's working power supply. The appearance of several common inverters is shown in Figure 4-1.

Figure 4-1 Appearance of several common inverters

The main function of the inverter is to change the power supply frequency of the motor to adjust the load, reduce power consumption, and energy loss, extend the service life of the equipment, and increase the degree of automation of the production equipment. At present, the frequency converter has been widely used in the production of light and heavy industry and people's daily life.

Inverter Structure and Classification

• **Structure of Inverter**

Inverter is composed of the main circuit and the control circuit. The main circuit consists of the rectifier circuit, the filter circuit and the inverter circuit. The structural block diagram of the inverter is shown in Figure 4-2.

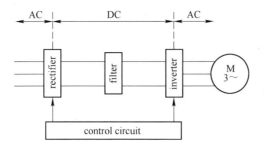

Figure 4-2 Structure block diagram of the inverter

(1) Main Circuit of Inverter: The rectifier circuit is composed of a three-phase full-wave rectifier bridge. Its function is to rectify an external power supply of power frequency, generate a pulsating DC voltage, and provide the DC power required for the inverter circuit and the control circuit. The filter circuit smoothes the output of the rectifier circuit to ensure that the inverter circuit and the control circuit can obtain a high-quality DC power supply. The inverter circuit is to convert the DC voltage source output from the DC intermediate circuit into an AC power source with adjustable frequency and voltage.

(2) Control Circuit of Inverter: The control circuit of the inverter is composed of an arithmetic circuit, a detection circuit, a control signal input and output circuit, and a drive circuit. Its main functions are to complete the switching control of the inverter circuit, the voltage control of the rectifier circuit, and to complete various protection functions.

- **Classification of Inverter**

The classifications of inverter are as follows:

(1) Classified by Input Voltage Level. Inverters can be divided into low-voltage inverters and high-voltage inverters according to the input voltage level. Common low-voltage inverters include single-phase 220V inverters, three-phase 220V inverters, and three-phase 380V inverters. High-voltage inverters commonly have 6kV, 10kV inverters.

(2) Classification by Frequency Conversion Method. The inverter is divided into AC-AC inverter and AC-DC-AC inverter according to the frequency conversion method.

The AC-AC inverter can directly convert the power frequency AC current into an AC current whose frequency and voltage can be controlled, so it is called a direct inverter. The maximum frequency of this kind of inverter output can only reach 1/3 ~ 1/2 of the power frequency, which is suitable for low frequency and large capacity speed regulation system.

AC-DC-AC inverter converts industrial frequency AC power into DC power through rectifier, and then converts DC power into AC power with adjustable frequency and voltage, so it is also called indirect inverter. At present, the widely used universal inverters are AC-DC-AC inverters.

(3) Classification by inverter purpose. According to the purpose of inverter, it can be divided into general inverter and special inverter. The general inverter is an inverter that can be applied to all loads. The special inverter is a series of special optimizations for the characteristics of the load dragged by the inverter on the basis of the general inverter. It has a series of advantages such as simpler parameter setting and better energy-saving speed regulation effect. Common special inverters include inverters for fans, inverters for pumps, inverters for elevators, etc.

4.1.3.2 Introduction of SINAMICS G120C Inverter

SINAMICS G120C inverter is an integrated compact inverter used to control the speed of three-phase AC motors produced by Siemens Company of Germany. It is one of the members of SINAMICS drive family. It can meet the needs of many applications such as conveyor belt, mixer, extruder, water pump, fan, compressor and some basic material processing machinery. The power range is from 0.55kW to 132kW, and it supports PROFINET, Ethernet/IP, PROFIBUS,

USS/Modbus RTU and other communication modes. The product can be built into the control box and switch cabinet, thereby saving space. Its appearance is shown in Figure 4-3.

Figure 4-3 Appearance of SINAMICS G120C inverter

SINAMICS G120C inverter consists of control unit (CU) and power module (PM). In the following, the functions and wiring methods of the terminals of these two parts are introduced respectively.

Terminal Introduction of SINAMICS G120C Inverter

The terminal functions and wiring methods of the control unit of SINAMICS G120C inverter are shown in Figure 4-4.

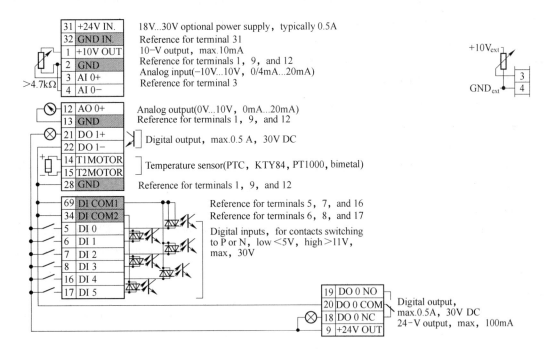

Figure 4-4 Terminals and wiring of SINAMICS G120C inverter

- **Power Terminal**

Terminals 1 (+10V OUT) and 2 (GND) are a high-precision 10V DC power supply provided by inverter for users.

Terminals 9 (+24V OUT) and 28 (GND) are internal 24V DC power supply of inverter, which can be used for digital input terminals. Terminal 9 (+24V OUT) is the positive terminal of 24V DC power supply inside the inverter, and the maximum output current is 10mA. Terminal 28 (GND) is the negative terminal of 24VDC power supply.

Terminals 31 (+24V IN) and 32 (GND IN) are externally connected 24V DC power supplies, and users provide 24V DC power supplies for the control end of the inverter.

- **Digital Terminal**

Terminals 5 (DI 0), 6 (DI 1), 7 (DI 2), 8 (DI 3), 16 (DI 4), 17 (DI 5) provide users with 6 fully programmable digital inputs, digital input signals Input the CPU through the photoelectric isolation to carry out forward/reverse rotation, forward/backward jog, fixed frequency set value control, etc.

Terminal 18 (DO 0 NC), 19 (DO 0 NO), 20 (DO 0 COM) and 21 (DO 1 POS), 22 (DO 1 NEG) digital output terminals, of which 18, 19, 20 are relay-type outputs; 21, 22 are transistor output.

- **Analog Terminal**

Terminals 3 (AI 0+) and 4 (AI 0-) provide users with analog voltage reference input, which can be used as frequency reference information to convert the analog value to digital value through the converter's internal analog/digital converter, transfer to the CPU.

Terminals 12 (AO 0+) and 13 (GND) are analog output terminals, which can provide standard DC analog signals for instrumentation or controller input terminals.

- **Communication Terminal**

The control unit provides users with two ProfiNet communication interfaces for data communication with other controllers.

- **Protection Terminal**

Terminals 14 (T1 MOTOR) and 15 (T2 MOTOR) are motor overheat protection input terminals. When the motor is overheated, it provides a trigger signal to the CPU.

- **Common Terminal**

Terminals 34 (DI COM2) and 69 (DI COM1) are digital common terminals. When using digital inputs, the corresponding common terminal must be connected to the negative terminal of the 24V power supply.

Practice

Observe the appearance of the G120C inverter and check the parameters on the product label on the right side of the inverter. Remove the basic operator panel of BOP-2, open the cover below the BOP-2, observe the position and label of the inverter's wiring terminals, corresponding to Figure 4-4, and think about how to wire.

Connection of SINAMICS G120C Inverter Power Unit

The connection between G120C inverter and input power supply and motor is shown in Figure 4-5. The input and output reactors and braking resistors in the figure are all optional components according to the needs of the inverter.

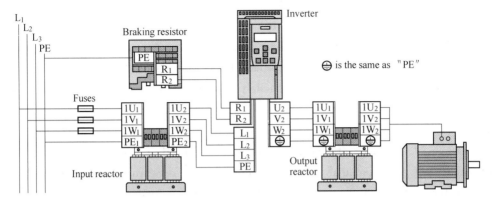

Figure 4-5　Connection of SINAMICS G120C inverter with power supply and motor

The functions and wiring of input and output reactors and braking resistors are as follows:

(1) Braking resistor: The braking resistor can effectively brake the high-inertia load for the inverter. Connect to the R_1 and R_2 terminals at the bottom of the G120C inverter, as shown in Figure 4-6.

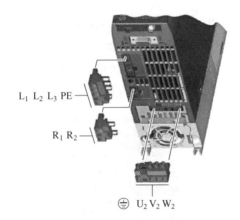

Figure 4-6　Interface terminals of SINAMICS G120C inverter and power supply and motor

(2) Input reactor: The input reactor can enhance the protection of the inverter against overvoltage, harmonics and commutation disturbances. Its input terminal is connected to L_1, L_2, L_3 of the three-phase power supply through fuses, as shown in Figure 4-5; The output terminal is connected to the L_1, L_2, L_3 and PE terminals located at the bottom of the G120C inverter, as shown in Figure 4-6.

(3) Output reactor: The output reactor is used to extend the effective transmission distance of the inverter and effectively suppress the instantaneous high voltage generated when the IGBT module of the inverter is switched. Its input end is connected to the U_2, V_2, W_2 and PE terminals at the bottom of the G120C inverter, as shown in Figure 4-6; The output end is connected to the

terminal of the three-phase AC motor, as shown in Figure 4-5.

Before installing the inverter, please refer to the product installation manual for the selection of the braking resistor, input reactor, output reactor, power cable, signal cable and the installation method of the inverter. During the installation process, the inverter must be reliably grounded, the power supply must be disconnected when the cable is connected or the wiring is changed, and the power and signal wires must be laid separately to prevent electromagnetic interference from affecting the normal operation of the device.

Practice

Connect the L_1, L_2, and L_3 three-phase power cables and PE ground cables of the G120C inverter to the input power terminal. Connect the three-phase asynchronous motor to the U_2, V_2, W_2 and PE terminals of the inverter. Pay attention to the wiring method of the three-phase asynchronous motor winding.

4.1.3.3 Commissioning of SINAMICS G120C Inverter

When using a new inverter for the first time or replacing the motor driven by the inverter, the nameplate parameter data of the motor and some basic drive control parameters need to be input into the inverter to better drive the motor. Below, the basic operator panel of BOP-2 is used to set and debug the inverter.

Cognition and Installation of Basic Operator Panel of BOP-2

The basic operator panel of BOP-2 is shown in Figure 4-7. It can be used for debugging, operation monitoring and setting of a certain parameter of the inverter. The menu navigating and two lines of parameters are used to display the debugging process of the inverter.

Figure 4-7 Basic operation panel of BOP-2
1—Menu bar indicating the currently selected menu function; 2—Provide information about the selected function or display the actual value; 3—Display the value

To plug basic operator panel of BOP-2 onto the inverter, the operation steps are shown in Figure 4-8.

To plug basic operator panel of BOP-2 onto the inverter, the operation steps are as follows:

(1) Remove the blanking cover of the inverter.

Figure 4-8 Installing basic operator panel of BOP-2

(2) Locate the lower edge of the BOP-2 housing in the matching recess of the inverter housing.

(3) Press the BOP-2 onto the inverter until you hear the latching mechanism on the inverter housing engage.

There are 7 function keys on BOP-2, and their functions are shown in Table 4-1.

Table 4-1 Functions of basic operation panel of BOP-2 keys

Key	Function
OK	The ⟨OK⟩ key has the following functions: (1) When navigating through the menus, pressing the key confirms selection of a menu item. (2) When working with parameters, pressing the OK key allows the parameter to be modified. Pressing the ⟨OK⟩ key again confirms the entered value and return you to the previous screen. (3) In the faults screen, the key is used to clear faults
▲	The ⟨UP⟩ key has the following functions: (1) When navigating a menu, the key moves the cursor up through the displayed list. (2) When editing a parameter value, the displayed value is increased. (3) If HAND mode is active and the jog function is active; A long press of the ⟨UP⟩ and ⟨DOWN⟩ key simultaneously has the following effects: 1) If reverse is ON, the ⟨UP⟩ key switches the reverse function OFF; 2) If reverse is OFF, the ⟨UP⟩ key switches the reverse function ON
▼	The ⟨DOWN⟩ key has the following functions: (1) When navigating a menu, the key moves the cursor down through the displayed list. (2) When editing a parameter value, the displayed value is decreased
ESC	The ⟨ESC⟩ key has the following functions: (1) If pressed for less than 2s, the BOP-2 returns to the previous screen. When a displayed value has been edited, the new value is not saved. (2) If pressed longer than 3s, the BOP-2 returns to the status screen. When using the ⟨ESC⟩ key in the parameter edit mode, no data is saved unless the ⟨OK⟩ key has been pressed

Continued Table 4-1

Key	Function
⏽ (ON)	The ⟨ON⟩ key has the following functions: (1) In AUTO mode, the key is not active and if pressed, the key is ignored. (2) In HAND mode the inverter starts the motor; The operator panel screen displays the drive running icon.
○ (OFF)	The ⟨OFF⟩ key has the following functions: (1) In AUTO mode the key has no effect and the key press is ignored. (2) If pressed for longer than 2s the inverter performs an OFF2; The motor coasts down to a standstill. (3) If pressed for less than 3s, the following actions will be performed: 1) If the ⟨OFF⟩ key is press twice in less than 2s, OFF2 is performed. 2) If in HAND mode, the Inverter performs an OFF1; The motor ramp-down to a standstill in the time set in parameter P1121
HAND/AUTO	The ⟨HAND/AUTO⟩ key switches the command source between the BOP-2 (HAND) and fieldbus (AUTO). The functions are as follows: (1) If HAND mode is active, the ⟨HAND/AUTO⟩ key switches the inverter to AUTO mode and disables the ⟨ON⟩ and ⟨OFF⟩ keys. (2) If AUTO mode is active, the ⟨HAND/AUTO⟩ key switches the Inverter to HAND mode and enables the ⟨ON⟩ and ⟨OFF⟩ keys. (3) Changing between HAND mode and AUTO mode is possible while the motor is still running

Practice

Practice the disassembly and installation of basic operation panel of BOP-2. Turn on the inverter power supply and become familiar with the functions of the panel keys.

Menu Structure and Functions of Basic Operation Panel of BOP-2

The menu structure and functions of the basic operation panel of BOP-2 are shown in Figure 4-9.

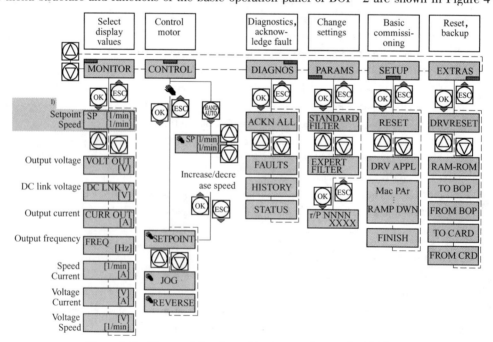

Figure 4-9 Menu and functions of basic operation panel of BOP-2

Table 4-2 shows a brief description of menu functions.

Table 4-2 Basic operation panel of BOP-2 menu function description

Menu	Functional description
MONITOR	Monitor menu: displaying the actual speed, actual output voltage and current value of the motor
CONTROL	Control menu: allowing the user to access the setpoint, jogging and reversing functions of the inverter
DIAGNOS	Diagnostics menu: displaying of fault alarm, control word and status word
PARAMS	Parameter menu: allowing access to view and changing the parameters of the Inverter
SETUP	Setup menu: allowing the user to perform the standard commissioning of the Inverter
EXTRAS	Extras menu: allowing users to restore factory default settings and data backups

The basic operation panel of BOP-2 will display various icons on the left side of the display to indicate the actual status of the inverter. The description of the icons is shown in Table 4-3.

Table 4-3 Basic operation panel of BOP-2 icon description

Icon	Features	Status	Remarks
	Command source	Hand	When the HAND mode is active, the icon is displayed; When AUTO mode is active, no icon is displayed
	Inverter status	Inverter and motor running	The icon is static and does not rotate
JOG	Jog function	Jog function active	The inverter jog function is activated
	Error/Alarm	Fault or alarm pending Flashing symbol = Fault Steady symbol = Alarm	If a fault is detected, the inverter will be stopped and the user is required to take the necessary corrective actions to clear the fault. An alarm is a condition that will not stop the inverter

Using Basic Operation Panel of BOP-2 to Modify Parameters

Modify the parameter value in the menu PARAMS, the specific steps are as follows:

(1) Select parameter number. When the displayed parameter number flashes, press the ⟨UP⟩ or the ⟨DOWN⟩ key to select the desired parameter number; Press the ⟨OK⟩ key to enter the parameter and display the current parameter value.

(2) Modify parameter value. When the displayed parameter value flashes, press the ⟨UP⟩ or the ⟨DOWN⟩ key to adjust the parameter value; Press the ⟨OK⟩ key to save the parameter value.

Practice

Power on the inverter, be familiar with the menu operation method of the panel, and practice modifying parameter values.

Using Basic Operation Panel of BOP-2 for Quick Commissioning of Inverter

The quick commissioning of inverter is an operation mode of simple and fast running motor by setting the basic parameters of motor. The preconditions for quick commissioning are as follows:

(1) The power supply is switched on.

(2) The operator panel displays setpoints and actual values.

Proceed as follows can carry out quick commissioning:

(1) Press the ⟨OK⟩ key.

(2) Press the ⟨UP⟩ key until BOP-2 displays the 'SETUP' menu.

(3) Press the ⟨OK⟩ key in the 'SETUP' menu, to start quick commissioning.

(4) If you wish to restore all of the parameters to the factory setting before the quick commissioning, the steps are proceed as follows:

1) Press the ⟨OK⟩ key.

2) Switchover the display using an arrow key: NO → YES.

3) Press the ⟨OK⟩ key.

The display shows BUSY, which changes to DONE after a while, and finally displays DRV_APPL P96_.

(5) When you select an application class, the inverter assigns suitable default settings to the motor control:

1) 0 EXPERT;

2) 1 STANDAR (Standard Drive Control);

3) 2 DYNAMIC (Dynamic Drive Control).

Select STANDAR. Refer to Table 4-4 for related parameter settings during the quick commissioning of the inverter.

Table 4-4 Parameter settings during quick commissioning of the inverter

Parameters	Functional	Settings
EUR/USA P100	Select the motor standard, the system provides 3 options: 0 kW 50Hz, 1 HP 60Hz, and 2kW 60Hz	kW 50Hz IEC
INV VOLT P210	Set the inverter supply voltage	380
MOT TYPE P300	Select the motor type	INDUCT (Third-party induction motor)
MOT CODE P301	Motor code: If you use the motor code to select the motor type, you need to enter; If you do not know the motor code, then you must set the motor code=0	0
87 HZ	87 Hz motor operation: The BOP-2 only displays this step if you previously selected IEC as the motor standard, Choose no or yes. 87Hz can make the running speed of the motor reach 1.73 times the normal speed	No
MOT VOLT P304	Rated motor voltage	380

Continued Table 4-4

Parameters	Functional	Settings
MOT CURR P305	Rated motor current	Set according to the rating plate of motor
MOT POW P307	Rated motor power	Set according to the rating plate of motor
MOT FREQ P310	Rated motor frequency	50
MOT RPM P311	Rated motor speed	Set according to the rating plate of motor
MOT COOL P335	Motor cooling, the system provides 4 options: (1) 0 SELE Natural cooling; (2) 1FORCED Forced-air cooling; (3) 2 LIAUID Liquid cooling; (4) 128 NO FAN Without fan	128 NO FAN
TEC APPL P501	Selecting the basic setting for the motor control, the system provides 2 options: (1) 0 VEC STD: Constant load; typical applications include conveyor drives. (2) 1 PUMP FAN: Speed-dependent load; typical applications include pumps and fans	0
MAc PAr P15	Select the default setting for the interfaces of the inverter that is suitable for your application	1
MIN RPM P1080	Minimum motor speed, the default value is 0	Set according to the rating plate of motor
MAX RPM P1082	Maximum motor speed, the default value is 1500r/min	Set according to the rating plate of motor
RAMP UP P1120	Ramp-up time of the motor, the default value is 10s	10
RAMP DWN P1121	Ramp-down time of the motor, the default value is 10s	10
FINISH	Ramp-down time after an OFF3 command, the default value is 0	0
MOT ID P1900	Motor data detection, select the way the inverter measures the data of the connected motor, the system provides 6 options: 0 OFF, 1 STIL ROT, 2 STIL, 3 ROT, 11 ST RT OP and 12 STILL OP	0 (Motor data is not measured)
FINISH	Complete quick commissioning	Switchover the display using an arrow key: NO→YES, Press the ⟨OK⟩ key

After successfully inputting all the data needed for quick commissioning of the inverter, the panel displays BUSY, the inverter calculates the parameters, and after the calculation is completed, the inverter displays the DONE interface, and the cursor returns to the MONITOR menu.

If P1900 is set to 0 during quick commissioning, the alarm A07991 will be displayed on the screen after quick commissioning, and the red fault LED will flash, prompting to activate motor data detection and waiting for the start command.

After the quick commissioning is completed, if there are other parameters that need to be modified, or some parameters of the motor in the quick commissioning need to be modified, they can be modified directly in the expert parameter list. To open the expert parameter list, first set the quick commissioning parameter P0010 = 1. Only when P0010 = 1, can the relevant parameters of the motor be modified. After the parameter modification is completed, P0010 = 0 must be made.

Practice

According to the nameplate data of the three-phase asynchronous motor connected to the inverter, use the basic operation panel of BOP-2 to set up the inverter for quick commissioning in HAND mode.

Changing settings using BOP-2

You can modify the settings of your inverter by changing the values of the its parameters. The parameters are divided into two types: writeable parameters and read-only parameters. The inverter only permits changes to writable parameters, writable parameters begin with a 'P', such as P45. Read-only parameters begin with an 'r', the value of a read-only parameter cannot be changed, such as r2.

Use BOP-2 to change the writable parameters according to the following steps. Take the modification of the parameter number P45 as an example, as shown in Figure 4-10.

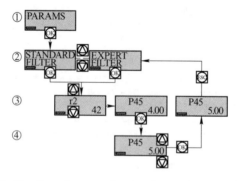

Figure 4-10 Operation steps for changing settings using BOP-2

As shown in Figure 4-10, the stepe are as follows:

(1) Select the PARAMS menu and press the ⟨OK⟩ key.

(2) Select the parameter filter using the arrow keys, press the ⟨OK⟩ key. STANDARD: The inverter only displays the most important parameters. EXPERT: The inverter displays all of the pa-

rameters.

(3) The last set parameter content is displayed on the screen. Select the required number of a writable parameter using the arrow keys, press the ⟨OK⟩ key, such as P45. At this time, the number '45' is in flashing state, press ⟨OK⟩ key, P45 content '4.00' starts flashing, indicating that it is in modifiable state.

(4) Select the value of the writable parameter using the arrow keys, such as '4.00' is changed to '5.00', after confirming the value, accept the value with press the ⟨OK⟩ key.

You have now changed a writable parameter using the BOP-2, and BUSY appeared on the screen. For indexed parameters, several parameter values are assigned to a parameter number, each of the parameter values has its own index.

Take the modification of the indexed parameter of P840 as an example, the operation steps are shown in Figure 4-11.

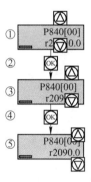

Figure 4-11 Operation steps for changing indexed parameters

As shown in Figure 4-11, the steps are as follows:

(1) After selecting parameter number P840, '840' flashes.

(2) After pressing the ⟨OK⟩ key, [00] begins to flash, indicating that this index can be modified.

(3) Use the arrow keys to set the parameter index, and you can change [00] to [01].

(4) Press the ⟨OK⟩ key.

(5) The indexed parameters were successfully changed.

4.1.3.4 Predefined Interface Macro of SINAMICS G120C Inverter

Overview of Predefined Interface Macros of G120C Inverter

SINAMICS G120C inverter integrates a variety of predefined interface macro functions, which can be directly called by the user. Each macro corresponds to a wiring method. After selecting one of the macros, the inverter will automatically set some parameters corresponding to its wiring method, which greatly facilitates the user's quick debugging and improves the debugging efficiency.

Modify the macro number through parameter P0015. Only when P0010 = 1, can the parameter value of P0015 be modified. The steps to modify P0015 are as follows:

(1) Set P0010=1;

(2) Modify the parameter value of P0015;

(3) Set P0010=0;

Table 4-5 shows 17 kinds of macro functions of G120C PN inverter.

Table 4-5 Macro functions of G120C PN inverter

Macro number	Macro function description
1	Two-wire control with two fixed speeds
2	Two fixed speeds in one direction with Safety function
3	Four fixed speeds in one direction
4	Fieldbus
5	Fieldbus with Safety function
6	Fieldbus with two safety function
7	Switchover between fieldbus and jogging
8	Motorized potentiometer (MOP) with safety function
9	Motorized potentiometer (MOP)
12	Two-wire control with method 1
13	Setpoint via analog input with safety function
14	Switchover between fieldbus and motorized potentiometer (MOP)
15	Switchover between analog setpoint and motorized potentiometer (MOP)
17	Two-wire control with method 2
18	Two-wire control with method 3
19	Three-wire control with method 1
20	Three-wire control with method 2

The following inverter controls the multi-speed operation of the motor, which uses the predefined interface macro with macro number 1. To adjust the motor speed by connecting an external potentiometer to the inverter, the predefined interface macro with macro number 13 is used.

Example of Predefined Interface Macro of G120C Inverter

• Predefined Interface Macro 1

Set G120C inverter parameter P0015=1, the function of the terminals is shown in Figure 4-12.

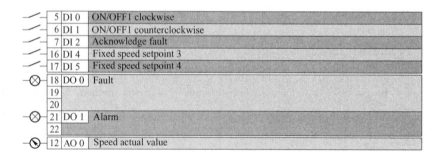

Figure 4-12 Predefined interface macro 1 terminal functions

The function of this macro is two-wire control with two fixed speeds. DI 0 is defined as the start/stop function of forward rotation; DI 1 is defined as the start/stop function of reverse rotation; DI 4 is defined as the fixed speed setpoint 3, which is set in P1003; DI 5 is defined as the fixed speed setpoint 4, The speed setpoint is set in P1004; when both DI 4 and DI 5 are hight, the inverter runs at 'fixed speed setpoint 3+fixed speed setpoint 4'.

- Predefined Interface Macro 13

Set G120C inverter parameter P0015 = 13, the function of the terminals is shown in Figure 4-13.

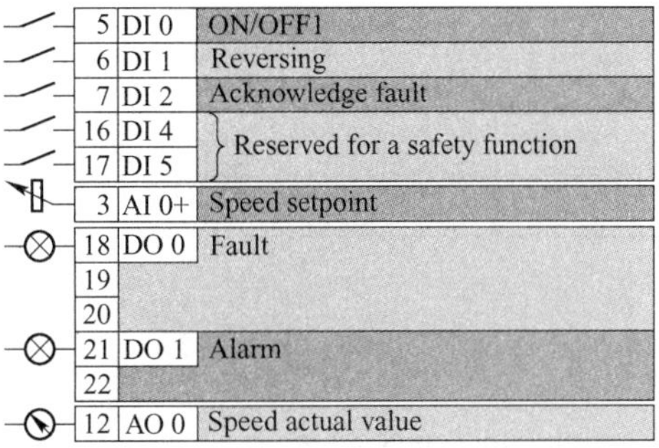

```
 —| 5 |DI 0 | ON/OFF1
 —| 6 |DI 1 | Reversing
 —| 7 |DI 2 | Acknowledge fault
 —|16 |DI 4 |}
 —|17 |DI 5 |  Reserved for a safety function
 —| 3 |AI 0+| Speed setpoint
⊗—|18 |DO 0 | Fault
   |19 |
   |20 |
⊗—|21 |DO 1 | Alarm
   |22 |
○—|12 |AO 0 | Speed actual value
```

图 4-13   Predefined interface macro 13 terminal functions

The function of this macro is setpoint via analog input with safety function. DI0 is defined as a start/stop function; DI1 is defined as a commutation function; AI0 + is defined as an external analog to adjust the speed setpoint.

4.1.3.5　SINAMICS G120C Inverter Command Source and Setpoint Source

Through the predefined interface macro, you can define what signal the inverter uses to control the start and what signal will control the output frequency.

Command Source

Command source refers to the interface where the inverter receives control commands. When setting the predefined interface macro P0015, the inverter will automatically define the command source. In the parameter settings listed in Table 4-6, r722.0, r722.2, r722.3, r2090.0, and r2090.1 are all command sources.

Table 4-6　Command source example

Par. No.	value	Description
P0840	722.0	Define DI 0 as start command
	2090.0	Define bit 0 of fieldbus control word 1 as the start command
P0844	722.2	Define DI 2 as OFF2
	2090.1	Define bit 1 of fieldbus control word 1 as the OFF2 command
P2013	722.3	Define DI 3 as fault reset

Setpoint Source

The setpoint source refers to the interface where the inverter receives the setpoint. When setting the predefined interface macro P0015, the inverter will automatically define the setpoint source. The common setpoint sources for the main setpoint P1070 are shown in Table 4-7. R1050, r755.0, r1024, r2050.1, and r755.1 are all setpoint sources.

Table 4-7 Setpoint source example

Par. No.	value	Description
P1070	1050	MOP as the main setpoint
	755.0	AI 0 as the main setpoint
	1024	Fixed speed as the main setpoint
	2050.1	Fieldbus process data as the main setpoin
	755.1	AI 1 as the main setpoint

4.1.4 Task Implementation

On the XK-SX2C advanced maintenance electrician training platform, a jumper is used to complete the connection between the inverter and the three-phase power supply, and the connection is completed between the inverter and the three-phase asynchronous motor. On the basis of making sure, the connection is correct, and the inverter is powered on. According to the rating plate data of the motor, the BOP-2 is used to complete the quick commissioning of the inverter, and on this basis, the following operating tasks are completed.

4.1.4.1 Operate Inverter BOP-2 to Control Start/Stop Operation of Motor

Task Requirements

Operate the BOP-2 panel to realize the start, stop and continuous operation of the three-phase asynchronous motor, adjust the motor speed, and then score the project after completion, as shown in Table 4-10.

Operation Steps

The operation steps are as follows:

(1) On the BOP-2, pressing the ⟨HAND/AUTO⟩ key, the HAND icon is displayed on the BOP-2 screen, and the inverter enters the HAND operation mode.

(2) Press the ⟨ESC⟩ key to enter the menu selection.

(3) Press the ⟨UP⟩ or ⟨DOWN⟩ key to move the menu bar to CONTROL and then press the OK key.

(4) Displaying SET POINT on the screen, and then pressing the ⟨OK⟩ key to display SP, the displayed motor speed value is 0.

(5) Press the ⟨UP⟩ key to adjust the motor speed setpoint, such as 1000r/min.

(6) Pressing the ⟨ON⟩ key to start the motor, the motor speed gradually rises to the setpoint.

(7) At this time, press the ⟨UP⟩ or ⟨DOWN⟩ keys to change the motor speed setpoint, thereby changing the motor speed.

(8) At this time, press the ⟨OK⟩ key to switch and display the voltage, current, frequency and other values of the motor operation.

(9) Pressing the ⟨OFF⟩ key, the motor stops running, and the speed gradually decreases to 0.

4.1.4.2 Operate Inverter BOP-2 to Control Jog Operation of Motor

Task Requirements

Operate BOP-2 to realize jog operation of three-phase asynchronous motor, and the score the project after completion, as shown in Table 4-10.

Operation Steps

The operation steps are as follows:

(1) On the BOP-2, press the ⟨HAND/AUTO⟩ key. The HAND icon is displayed on the BOP-2 screen, and the inverter enters the HAND operation mode.

(2) Press the ⟨ESC⟩ key to enter the menu selection.

(3) Press the ⟨UP⟩ or ⟨DOWN⟩ key to move the menu bar to CONTROL and then press the ⟨OK⟩ key.

(4) Displaying SET POINT on the screen, and then pressing the ⟨OK⟩ key to display SP, the displayed motor speed setpoint is 0.

(5) Press the ⟨UP⟩ key to set the motor speed setpoint, such as 1000r/min.

(6) Press the ⟨ESC⟩ key to return to the screen and display SET POINT.

(7) Pressing the ⟨DOWN⟩ key, the screen displays JOG. Then pressing the ⟨OK⟩ key, the screen displays OFF. Pressing the ⟨DOWN⟩ key, displaying On, and then pressing the ⟨OK⟩ key, the setting is successful. JOG is displayed in the lower left corner of the screen, and the motor operation is selected as the jog mode.

(8) Pressing and holding the ⟨ON⟩ key to start the motor, the motor speed gradually rises to a certain speed value, but it does not reach 1000r/min, because the motor speed of the jog operation is not set by SET POINT. The value of the jog speed is in the parameter No. P1058 is set, and the factory default is 150r/min.

(9) When the ⟨ON⟩ key is released, the motor stops running, and the speed gradually decreases to 0.

4.1.4.3 Operate Inverter BOP-2 to Control Reverse Running of Motor

Task Requirements

Operate BOP-2 panel to realize reverse running of three-phase asynchronous motor, and the score the project after completion, as shown in Table 4-10.

Operation Steps

The operation steps are as follows:

(1) On the BOP-2, pressing the ⟨HAND/AUTO⟩ key, the HAND icon is displayed on the BOP-2 screen, and the inverter enters the HAND operation mode.

(2) Press the ⟨ESC⟩ key to enter the menu selection.

(3) Press the ⟨UP⟩ or ⟨DOWN⟩ key to move the menu bar to CONTROL and then press the ⟨OK⟩ key.

(4) Displaying SET POINT on the screen, and then pressing the ⟨OK⟩ key to display SP, the displayed motor speed value is 0.

(5) Press the ⟨UP⟩ key to set the motor speed setpoint, such as 1000r/min.

(6) Pressing the ⟨ON⟩ key to start the motor, the motor speed gradually rises to the setpoint, observing and remembering the direction of the motor running.

(7) Press the ⟨ESC⟩ key to return to the screen and display SET POINT.

(8) Pressing the ⟨DOWN⟩ key twice, the screen displays REVERSE. Then pressing the ⟨OK⟩ key, the screen displays OFF. Pressing the ⟨DOWN⟩ key, displaying On, and then pressing the ⟨OK⟩ key, the setting is successful, and the motor operation is selected as the reverse mode.

(9) The motor gradually decelerates to 0 in the original running direction. At this time, the motor starts to reverse direction until it accelerates to the speed setpoint.

(10) Pressing the ⟨OFF⟩ key, the motor stops running, and the speed gradually decreases to 0.

Another operation procedures to change the running direction of the three-phase asynchronous motor are as follows:

(1) When the motor is running, the SET POINT menu is displayed on the screen; Then the OK key is pressed again to display the value of SP.

(2) If the rotation speed setpoint displayed by SP is a positive value, press and hold the ⟨DOWN⟩ key to decrease the rotation speed setpoint. When the setpoint shows a negative sign, the motor's steering will become reverse.

(3) Similarly, when the rotation speed setpoint displayed by SP is negative, press and hold the ⟨UP⟩ key. When the negative sign of the set value disappears and becomes a positive value, the motor becomes forward running.

The above three operation tasks are to use the BOP-2 to complete the simple setting and operation control of the inverter. The following operation tasks will use digital terminals to connect multiple switches to achieve multi-speed operation of three-phase asynchronous motors.

4.1.4.4 Inverter Controls Multisection Speed Operation of Motor

Task Requirements

Multisection speed operation is also called fixed frequency operation. By externally connecting a switch to the digital terminal of the G120C inverter, the motor can be operated at three speeds of 400r/min, 800r/min, and 1200r/min, respectively, and it can be switched between the three speeds through the switch.

Finally, score the project after completion, as shown in Table 4-10.

Task Analysis

Under the condition of setting P1000=3, through the external switch of the digital terminal of the G120C inverter, the combination of fixed setpoint is selected to realize the multisection speed operation of the motor. There are two fixed setpoint modes: direct selection and binary selection. The digital input terminals 5, 6, 7, 8, 16, and 17 of G120C inverter provide users with 6 fully programmable digital input terminals. The input signals can come from the output signals of external buttons, relays or transistors. Terminals 9 and 28 are the 24V DC power supply built into the inverter and provide the DC power required for digital input. By connecting a button to the digital input terminal and setting related parameters, the motor can be operated at multisection speeds and the start/stop control can be achieved.

Operation Steps

The operation steps are as follows:

(1) Inverter wiring: According to Figure 4-14, the wiring work is completed between the inverter and the digital terminals.

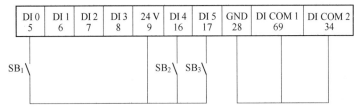

Figure 4-14 Wiring of digital terminal of G120C inverter

In Figure 4-14, three buttons are externally connected to the digital input terminals 5, 16, and 17. These buttons can be defined as functions such as forward start and stop, fixed frequency operation control, etc. through parameter setting.

(2) Inverter parameter setting: Refer to Table 4-8 for the parameter settings of the inverter to realize the motor running at three speeds of 400r/min, 800r/min and 1200r/min.

Table 4-8 Inverter parameter setting table

Par. No.	Parameter value	Features	Remarks
P0015	1	Predefined macro parameters. The macro number is 1, the macro function is two-wire control, and there are two fixed frequencies Note: When modifying P0015, it must be changed when P0010 = 1. After the modification, P0010 = 0	
P0840	722.0	With DI0 as the starting signal, r722.0 is the parameter of DI 0 state, and SB_1 is defined as the motor's forward rotation start/stop control function	
P1000	3	P1000 is the frequency control source parameter, and P1000 = 3 is fixed frequency operation	
P1016	1	P1016 = 1 is defined as the direct selection mode for fixed frequency operation. In this mode, P1003 is a fixed speed 3, and P1004 is a fixed speed 4. When both DI 4 and DI 5 are high, the inverter runs at the speed of 'fixed speed 3 + fixed speed 4'	Defaults
P1022	722.4	Using DI 4 as the selection signal of fixed speed 3, r722.4 is the parameter of DI 4 state, and SB_2 is the input of a selection signal	
P1023	722.5	Using DI 5 as the selection signal of fixed speed 4, r722.5 is the parameter of DI 5 state, and SB_3 is the input of a selection signal	

Continued Table 4-8

Par. No.	Parameter value	Features	Remarks
P1003	400	Defining a fixed speed of 3, the unit is r/min. Here set to 400r/min	Set up
P1004	800	Defining a fixed speed of 4, the unit is r/min. Here set to 800r/min	
P1070	1024	Define a fixed speed value as the main set value	Defaults

According to the relevant parameter settings in Table 4-8, SB_1 is defined as the control button for the motor to start and stop, SB_2 is defined as the control button that runs at 400r/min, and SB_3 is defined as the control that runs at 800r/min. Pressing the SB_2 and SB_3 buttons at the same time, the motor runs at 1200r/min.

Task Presentation

The task presentation is as follows:

(1) Pressing and holding SB_1, the motor is power on.

(2) Pressing SB_2, the motor runs at 400r/min; Releasing SB_2, the motor stops running.

(3) Pressing SB_3, the motor runs at 800r/min; Releasing SB_3, the motor stops running.

(4) Pressing SB_2 and SB_3 at the same time, the motor runs at 1200r/min; Releasing SB_2, SB_3, the motor stops running.

(5) Releasing SB_1, the motor is disconnected from the power supply and stops running.

The multisection speed operation of the three-phase asynchronous motor is realized by the external switch of the digital input terminal of the inverter, but the speed of the motor can only be switched between several speeds, and the continuous change of the speed cannot be achieved. In the following, the continuous adjustment of the motor speed is achieved by means of an external potentiometer connected to the analog input of the inverter.

4.1.4.5 G120C Inverter External Potentiometer to Adjust the Speed of Motor

Task Requirements

Use the external potentiometer on the G120C inverter to realize real-time continuous adjustment of the running speed of the three-phase asynchronous motor, and then score the project after completion, as shown in Table 4-10.

Task Analysis

The G120C inverter uses an externally connected potentiometer, and the generated analog voltage signal is sent to the analog input terminal 3 (AI 0+) as frequency reference information. The input analog value is converted into a digital value by the internal analog/digital converter of the inverter and transmitted to the CPU adjusting the frequency of the inverter's output power. The terminals 1 and 2 of the G120C inverter are the 10V DC power supply built into the inverter, which is used as the DC power supply required by the potentiometer. By adjusting the resistance of the potentiometer, the voltage value can be changed continuously, and the running speed of the motor can be adjusted continuously.

Operation Steps

The operation steps are as follows:

(1) Inverter wiring: the wiring of the analog and digital terminals of the inverter is completed, as shown in Figure 4-15.

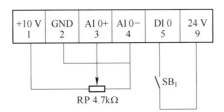

Figure 4-15 Inverter terminal wiring

As shown in Figure 4-15, the external button SB_1 is connected to the digital input terminal 5 through parameter setting, which is defined as the forward start and stop function. Wire-wound potentiometer RP with a resistance of 4.7kΩ, the DC voltage at both ends is taken from the 10V DC power supply inside the G120C inverter. The adjustable end of the potentiometer is connected to terminal 3 of the inverter, and terminal 4 is connected to terminal 2 GND. In addition, a DIP switch is setted behind the protective cover on the front of the inverter control unit to the 'U' position, indicating that the voltage input signal is used.

(2) Inverter parameter setting: Refer to Table 4-9 for the parameter settings of the inverter that realizes continuous motor speed adjustment.

Table 4-9 Inverter parameter setting table

Par. No.	Parameter value	Features
P0015	13	The interface macro parameters are predefined. The macro number is 13, and the macro function for the terminal start is given by the analog quantity Note: When modifying P0015, it must be changed when P0010 = 1. After the modification, P0010 = 0
P0840	722.0	Using DI 0 as the starting signal, r722.0 is the parameter of DI 0 state, and SB_1 is defined as the motor's forward rotation start/stop control function
P0756	0	Selection of analog input type: (1) 0: 0~10V; (2) 1: 2~10V; (3) 2: 0~20mA; (4) 3: 4~20mA; (5) 4: -10~+10V; (6) 8: No sensor is connected, choosing 0
P0757	0	0V corresponds to a frequency of 0Hz, that is, the corresponding motor speed is 0 r/min
P0758	0	
P0759	10	10V corresponds to a frequency of 50Hz, that is, the speed of the corresponding motor is the rated speed
P0760	100	
P1080	400	Set the minimum speed of the motor to 400 r/min
P1082	1300	Set the maximum speed of the motor to 1300 r/min

According to the relevant parameter settings of Table 4-9, SB_1 is defined as the motor start/stop control button. After the motor starts running, the speed is continuously adjusted between 400r/min and 1300r/min.

Task Presentation

The task presentation is as follows:

(1) Pressing and hold SB_1, the motor is turned on, and the motor starts to run.

(2) Adjust the potentiometer knob and observe the speed change of the motor. Using the MINITOR menu to observe the change range of the output speed of the inverter, it can be seen that the speed is continuously adjusted from 400r/min to 1300r/min.

(3) When SB_1 is released, the motor gradually stops.

4.1.5 Task Evaluation

Score the completion of the multisection speed operation task of the three-phase asynchronous motor controlled by the inverter, and fill the scoring results in Table 4-10.

Table 4-10 Score table

Task content	Assessment requirements	Grading	Distribution	Deduction	Score
Circuit design	Correctly design the wiring diagram of inverter control circuit	(1) The electrical control principle design function is incomplete, and 5 points are deducted for each missing function; (2) If the wiring diagram is incorrectly expressed or the drawing is not standardized, 2 points will be deducted from each place	10		
Circuit wiring	Inverter and three-phase power supply; wiring with three-phase asynchronous motor; wiring with digital terminals	(1) The connection between the inverter and the three-phase power supply and the three-phase asynchronous motor is deducted by 10 points for each wrong connection; (2) The connection between the digital terminal of the inverter and the button is deducted 5 points for each wrong connection	30		
Inverter parameter setting and debugging	Set the corresponding parameters of the inverter according to the control requirements specified in the task, and run and debug to meet the task requirements	(1) 10 points will be deducted each time when the inverter fails to run with power on; (2) 3 points will be deducted for each place when incompleting parameter setting of the inverter; (3) 2 points will be deducted for each place for wrong parameter setting; (4) 20 points will be deducted for no setting; (5) 5 points will be deducted each time when incorrecting operation of the inverter	50		

Continued Table 4-10

Task content	Assessment requirements	Grading	Distribution	Deduction	Score
Safe and civilized production	Abide by the 8S management system and the safety management system	(1) 10 points will be deducted for not wearing labor protection articles; (2) 5 points will be deducted each time when there are hidden safety hazards in the operation, until the deduction is completed. (3) 2 points will be deducted each time when the operation site is not sorted out and rectified in time until the deduction is completed	10		
		Total score			

4.1.6 Task Summary

The task implementation process record sheet is shown in Table 4-11.

Table 4-11 Task implementation process record sheet

	Task implementation process record sheet	
Fault 1	Fault phenomenon	
	Cause of failure	
	Troubleshooting process	
Fault 2	Fault phenomenon	
	Cause of failure	
	Troubleshooting process	
Fault 3	Fault phenomenon	
	Cause of failure	
	Troubleshooting process	
	Summary	

4.1.7 Task Development

Task requirements: The digital output terminal of SINAMICS G120C inverter can be used to indicate the operating status of three-phase asynchronous motor, such as digital output DO 0. The terminal number is 18, 19, and 20, which belongs to the relay type output. The corresponding parameter number is P0730. A 24V DC power supply and indicator light are connected between terminals 19 and 20 to show the running status of the motor.

On the basis of the multisection speed operation tasks of the inverter controlling the three-phase asynchronous motor, the product manual is queried by itself, the circuit is redesigned, the wiring is completed, the related parameter settings are completed, and the indication of the motor running state is realized.

Task 4.2 Using S7-1215C PLC and Inverter to Control Operation of Three-phase Asynchronous Motor

4.2.1 Task Description

As an alternative to traditional relays, PLC has been widely used in various fields of industrial control. It can change the control process through software, and has the advantages of small size, flexible assembly, simple programming, strong anti-interference ability and high reliability. It is also very suitable for applications in harsh working environments. When using an inverter to form an automatic control system for control, many cases are used in conjunction with PLC.

This task uses Siemens S7-1215C PLC and SINAMICS G120C inverter to form a control system to complete the control of the start and stop, forwarding and reversing rotation and multisection speed operation of the three-phase asynchronous motor.

4.2.2 Task Target

(1) Master the wiring method of PLC controlling Siemens G120 inverter.

(2) Master the setting of common parameters of Siemens G120C inverter.

(3) Able to independently complete the design, installation, programming, debugging and operation of the start/stop control circuit of three-phase asynchronous motor using Siemens S7-1215C PLC and G120 inverter.

(4) Can independently complete the design, installation, programming, debugging and operation of the three-phase asynchronous motor's forward and reverse control circuit using Siemens S7-1215C PLC and G120 inverter.

(5) Can independently complete the design, installation, programming, debugging and operation of the multisection speed control circuit of the three-phase asynchronous motor using Siemens S7-1215C PLC and G120 inverter.

4.2.3 Task-related Knowledge

4.2.3.1 PLC and Inverter to Realize Start/Stop Control of Three-phase Asynchronous Motor

The control unit of the Siemens inverter G120C PN integrates a variety of predefined interface macros, which can be directly called by the user, thereby improving the debugging efficiency. Using 2 fixed frequency transmission technologies, The corresponding parameter of the inverter is $P0015 = 1$. The meaning of the terminals corresponding to this scheme is shown in Figure 4-12.

The start and stop of the motor is controlled by digital input DI 0. The motor starts when DI0 is

1, and stops when DI 0 is 0.

Through digital input selection, two fixed speeds can be set. In this task, only one fixed speed is set. When digital input DI 4 is turned on, fixed speed 3 is used, which is set by P1003 parameter.

The input of S7-1215C PLC is controlled by buttons. There are two buttons for start and stop, output Q0.0 as the start signal of the motor, connect to the terminal 5 of the inverter (DI 0); Q0.1 connect to the terminal 16 of the inverter (DI 4). When both Q0.0 and Q0.1 outputs are 1, the inverter is set to a fixed speed output, such as the motor running at 1000r/min.

Practices

(1) How to realize the start/stop control of three-phase asynchronous motor? Fill the corresponding settings of inverter DI 0 and DI 4 into Table 4-12.

Table 4-12 DI 0 and DI 4 corresponding settings

DI 0	DI 4	Meaning

(2) Determine the correspondence between PLC and inverter terminals according to the above analysis, and fill in Table 4-13.

Table 4-13 Correspondence between PLC and inverter terminals

PLC				Inverter		
Input			Output	Input		
Function	Element	Address	Address	No. Terminal	Function	Meaning
Start						
Stop						

4.2.3.2 PLC and Inverter to Realize Forward and Reverse Control of Three-phase Asynchronous Motor

A variety of methods can be used to realize the forward and reverse control of the motor. In this example, the G120C PN inverter pre-defined interface macro 1 is used to implement, that is, the use of 2 fixed frequency transmission technologies. The coresponding parameter of the inverter is P0015=1.

The forward start and stop of the motor is controlled by digital input DI 0; The reverse start and stop of the motor is controlled by digital input DI 1.

Through digital input selection, two fixed speeds can be set. In this task, only one fixed speed is set. When digital input DI 4 is turned on, fixed speed 3 is used, which is set by P1003 parameter.

The input of the S7-1215C PLC is controlled by two buttons, start and stop. Output Q0.0 is the forward start signal of the motor, connected to the terminal 5 of the inverter; Output Q0.2 is the reverse start signal of the motor, connected to the terminal 6 of thes inverter; Q0.1 is connected to the terminal 16 of the inverter.

Practices

(1) How to realize forward and reverse control of three-phase asynchronous motor? Fill in the corresponding settings of the inverter DI 0, DI 1, and DI 4 in Table 4-14.

Table 4-14 DI 0, DI 1, and DI4 corresponding settings

DI 0	DI 1	DI 4	Meaning

(2) Determine the correspondence between PLC and inverter terminals according to the above analysis, and fill in Table 4-15.

Table 4-15 Correspondence between PLC and inverter terminals

PLC				Inverter		
Input			Output	Input		
Function	Element	Address	Address	No. Terminal	Function	Meaning

4.2.3.3 PLC and Inverter Realize Multisection Speed Control of Three-phase Asynchronous Motor

A variety of methods can be used to achieve the multisection speed control of the motor. In this example, the G120C PN inverter pre-defined interface macro 1 is used to implement, that is, the use of 2 fixed frequency transmission technologies. The corresponding parameter of the inverter is P0015 = 1.

The start and stop of the motor are controlled by digital input DI 0.

Through digital input selection, two fixed speeds can be set. When digital input DI 4 is turned on, fixed speed 3 is used; When digital input DI 5 is turned on, fixed speed 4 is used. The P1003 parameter sets a fixed speed of 3, and the P1004 parameter sets a fixed speed of 4. When DI 4 and DI 5 are connected at the same time, the speed of the inverter is the sum of fixed speed 3 and fixed speed 4.

The input of S7-1215C PLC is controlled by the start and stop buttons. Output Q0.0 is the starting signal of the motor, connected to the terminal 5 (DI 0) of the inverter. Q0.1 and Q0.2 are respectively connected to terminals 16 (DI 4) and 17 (DI 5) of the inverter. When both Q0.0

and Q0.1 outputs are 1, and the inverter is set as the first-section speed output, such as the motor running at 800 r/min. When both Q0.0 and Q0.2 outputs are 1, the inverter is set as the second-section speed output, such as the motor running at 400r/min. When the Q0.0, Q0.1 and Q0.2 outputs are all 1, the inverter is set as the third-section speed output, such as the motor running at 1200r/min.

Practices

(1) How to realize three-section speed control of three-phase asynchronous motor? Fill the corresponding settings of the inverter DI 0, DI 4, and DI 5 into Table 4-16.

Table 4-16　DI 0, DI 4, and DI 5 corresponding settings

DI 0	DI 4	DI 5	Meaning

(2) Determine the correspondence between PLC and inverter terminals according to the above analysis, and fill in Table 4-17.

Table 4-17　Correspondence between PLC and inverter terminals

PLC				Inverter			
Input		Output	Input				
Function	Element	Address	Address	No. Terminal	Function	Meaning	

4.2.4　Task Implementation

On the XK-SX2C advanced maintenance electrician training platform, the following three tasks is completed: PLC and inverter to achieve start and stop control, forward and reverse control and multisection speed control of three-phase asynchronous motor. Each task is required to connect the power supply on the basis of ensuring that PLC and inverter are connected correctly. Pay attention to the development of core qualities such as safety awareness and team awareness.

4.2.4.1　PLC and Inverter to Realize Start and Stop Control of Three-phase Asynchronous Motor

Task Requirements

Through the output of S7-1215C PLC and the digital input terminal of G120C PN inverter, the start and stop control of the motor is realized.

Finally, score the project after completion, as shown in Table 4-24.

Operation Steps

The operation steps are as follows:

(1) Wiring. The wiring work is completed between PLC and inverter as shown in Figure 4-16.

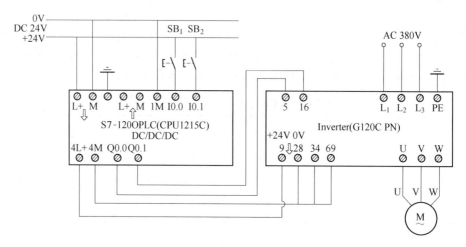

Figure 4-16 Wiring diagram of forward start-stop control system

(2) Inverter parameter setting. The steps are as follows:

1) Basic commissioning of the inverter is required.

2) The motor parameters is setted according to the actual motor parameters, such as:

① [MOT VOLT P304] [OK]   Rated voltage 380V;

② [MOT CURR P305] [OK]   Rated current 1.12A;

③ [MOT POW P307] [OK]   Rated power 0.180kW;

④ [MOT FREQ P310] [OK]   Rated frequency 50Hz;

⑤ [MOT RPM P311] [OK]   Rated speed 1430r/min.

3) Select a predefined interface macro, such as:

[MAc PAr P15] [OK]   P0015 = 1.

4) Motor speed related settings, such as:

① [MIN RPM P1080] [OK]   The minimum speed of the motor is set to 0;

② [MAX RPM P1082] [OK]   The maximum speed of the motor is set to 1430;

③ [RAMP UP P1120] [OK]   Motor acceleration time is set to 0.005;

④ [RAMP DWN P1121] [OK]   Motor deceleration time is set to 0.005;

⑤ [FINISH] [OK]   End basic commissioning: Use the arrow keys to switch NO-YES, and then press 〈OK〉 key.

The inverter parameters is setted according to Table 4-18.

Table 4-18 Parameter settings

Par. No.	Parameter value	Features
P0015	1	Predefined macro parameter setting 1
P0840	722.0	Using DI 0 as the start signal, r722.0 is the parameter of DI 0 state
P1000	3	Fixed frequency operation
P1016	1	Fixed speed mode adopts direct selection
P1022	722.4	Using DI 4 as a new signal for the selection of fixed settings, r722.4 is the parameter of DI 4 state
P1003	1000	Defining a fixed speed 3, unit is r/min
P1070	1024	Defining a fixed speed value as the main set value

(3) PLC. Hardware configuration refers to adding new equipment, selecting the corresponding PLC model of the training platform and setting the corresponding IP address. The operation steps are as follows:

1) Open the TIA Portal software and create an S7-1200 project.

2) Opening the project view, and clicking 'Add New Device', the Add New Device dialog box pops up, then selecting the controller. The specific path is as follows: SIMATIC S7-1200-> CPU->CPU 1215C DC/DC/DC->6ES7 215-1AG40-0XB0, as shown in Figure 4-17. The added PLC is shown in Figure 4-18.

3) Double click the Ethernet interface in Figure 4-18 and set the PLC's IP address to 192.168.0.1, as shown in Figure 4-19.

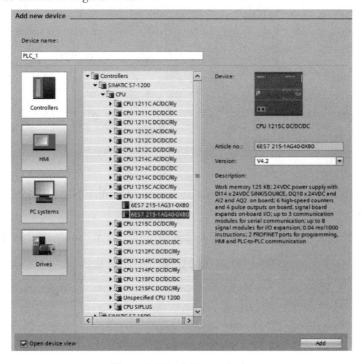

Figure 4-17 Add new device

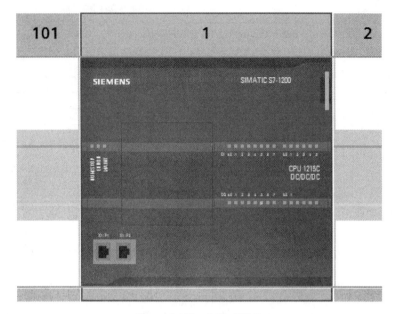

Figure 4-18 Added PLC

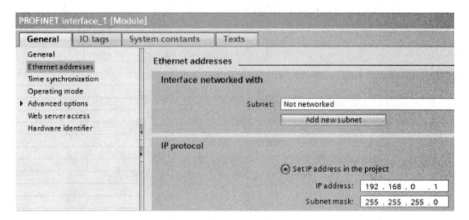

Figure 4-19 Setting the PLC's IP address

Software programming refers to writting programs, downloading and debugging. The operation steps are as follows:

1) Determine the I/O address allocation according to Figure 4-16 and fill in Table 4-19;

Table 4-19 I/O address allocation table

Input signal				Output signal		
Serial number	Function	Element	Address	Serial number	Inverter terminal name	Address

2) Write a program and download it to the PLC for debugging. After the device is configured, the program is written in the program block. Simple programs can be directly written to Main [OB1], or new blocks are added. Programs are written in the new blocks, and they are called through OB1. In this example, write the program directly in OB1, and double click (Main [OB1]) to enter the screen, as shown in Figure 4-20. The reference program is shown in Figure 4-21.

After the program editing is completed, download it to the PLC and monitor the program whether the running status meets the requirements. Pressing the start button, the monitoring result is shown in Figure 4-22. Pressing the stop button, the monitoring result is shown in Figure 4-23.

3) Connect the three-phase asynchronous motor to the line, and observe the running status after power on.

4) Summarize and record the data.

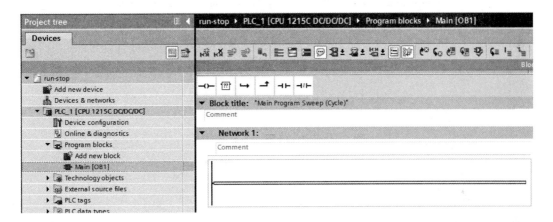

Figure 4-20　Programming entry OB1

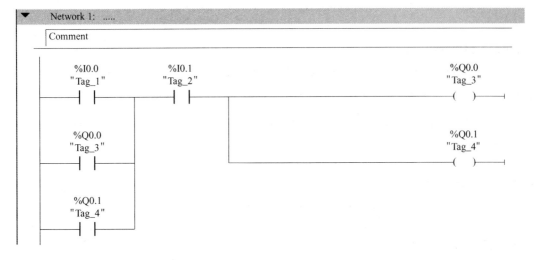

Figure 4-21　Reference program

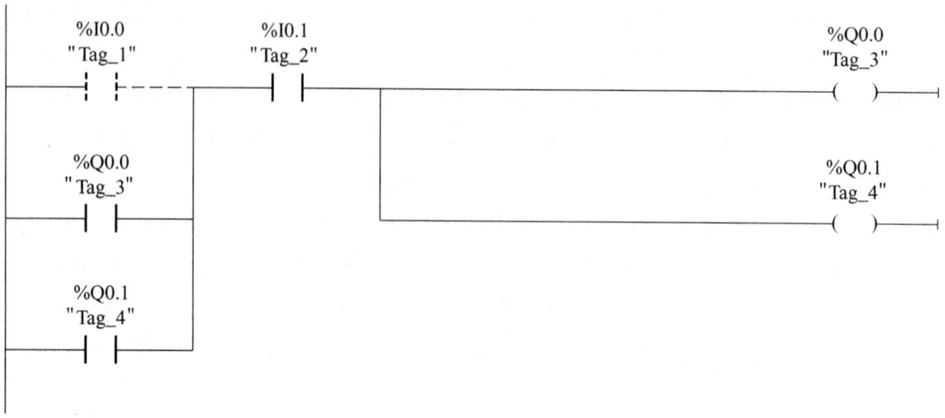

```
    %I0.0      %I0.1                              %Q0.0
    "Tag_1"    "Tag_2"                            "Tag_3"
    ─┤ ├────┬──┤ ├───┬─────────────────────────────( )────
             │       │
    %Q0.0    │       │                            %Q0.1
    "Tag_3"  │       │                            "Tag_4"
    ─┤ ├─────┤       └─────────────────────────────( )────
             │
    %Q0.1    │
    "Tag_4"  │
    ─┤ ├─────┘
```

Figure 4-22 Pressing the start button to monitor the program results

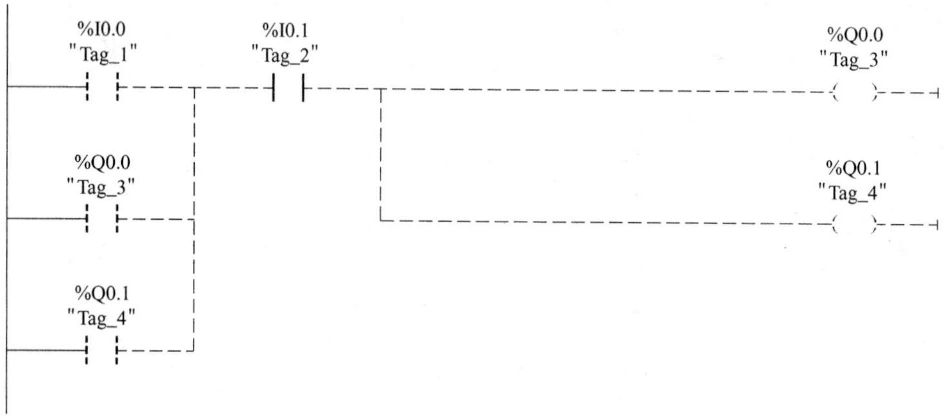

```
    %I0.0      %I0.1                              %Q0.0
    "Tag_1"    "Tag_2"                            "Tag_3"
    ─┤ ├────┬──┤ ├───┬─────────────────────────────( )────
             │       │
    %Q0.0    │       │                            %Q0.1
    "Tag_3"  │       │                            "Tag_4"
    ─┤ ├─────┤       └─────────────────────────────( )────
             │
    %Q0.1    │
    "Tag_4"  │
    ─┤ ├─────┘
```

Figure 4-23 Pressing the stop button to monitor the result of the program

Analysis and Summary

Summarize the task implementation process, find deficiencies, especially attach importance to the occurrence of malfunctions, and record in Table 4-25 after analysis and summary.

4.2.4.2 PLC and Inverter to Realize Forward and Reverse Control of Three-phase Asynchronous Motor

Task Requirements

Through the output of the S7-1215C PLC and the digital input terminal of the G120C PN inverter, the forward and reverse control of the motor is realized. Finally, scord the project after completion, as shown in Table 4-24.

The specific control requirements are as follows:

(1) When the start button is pressed, the Q0.0 and Q0.1 outputs of the PLC are both 1, and the motor rotates forward and runs at a speed of 1000r/min.

(2) After 5s, the Q0.2 and Q0.1 outputs of the PLC are both 1, and the motor reverses and runs at a speed of 1000r/min.

(3) Whenever the stop button is pressed, the motor stops immediately.

Operation Steps

The operation steps are as follows:

(1) Wiring. The wiring work is completed between PLC and inverter, as shown in Figure 4-24.

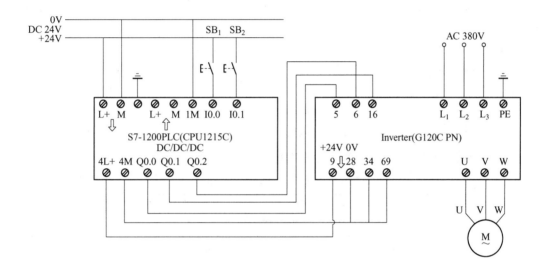

Figure 4-24 Wiring diagram of forward and reverse control system

(2) Inverter parameter setting. The steps are as follows:

1) The basic commissioning of the inverter is required.

2) Select the motor parameters according to the actual motor parameters.

3) Select the predefined interface macro.

4) Motor speed related settings.

The inverter parameters is setted according to Table 4-20.

Table 4-20 Parameter settings

Par. No.	Parameter value	Features
P0015	1	Predefined macro parameter setting 1
P0840	r722.0	Using DI 0 as the start signal, r722.0 is the parameter of DI 0 state
P1113	r722.1	Using DI 1 as the reverse signal, r722.1 is the parameter of DI 1 state
P1000	3	Fixed frequency operation

Continued Table 4-20

Par. No.	Parameter value	Features
P1016	1	Fixed speed mode adopts direct selection
P1022	722.4	Using DI 4 as a new signal for the selection of fixed settings, r722.4 is the parameter of DI 4 state
P1003	1000	Defining a fixed speed 3, unit is r/min
P1070	1024	Define a fixed speed value as the main set value

(3) PLC. Hardware configuration refers to adding new equipment, selecting the corresponding PLC model of the training platform and setting the corresponding IP address. The operation steps are as follows:

1) Open the TIA Portal software and create an S7-1200 project.

2) Opening the project view, and clicking 'Add New Device', the Add New Device dialog box pops up, the selecting CPU 1215 DC/DC/DC (6ES7 215-1AG40-0XB0).

3) Set the PLC's IP address to 192.168.0.1.

Software programming: Determine the I/O address allocation according to Figure 4-17, fill in Table 4-21, and write the program according to the requirements.

Table 4-21 I/O address allocation table

Input signal				Output signal		
Serial number	Function	Element	Address	Function	Inverter terminal name	Address

4) Write program and download to PLC for debugging. The reference program is shown in Figure 4-25.

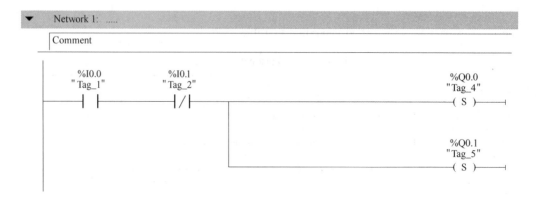

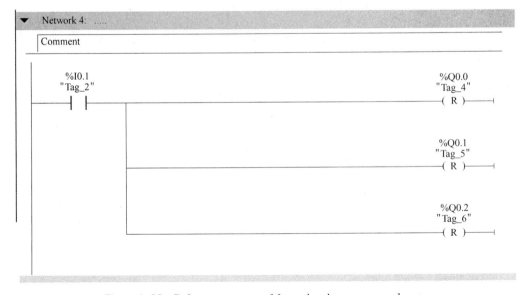

Figure 4-25 Reference program of forward and reverse control system

5) Connect the three-phase asynchronous motor to the line and observe the operating status after power-on.

6) Summarize and record the data.

Analysis and Summary

Summarize the task implementation process, find deficiencies, especially attach importance to the occurrence of malfunctions, and record in Table 4-25 after analysis and summary.

4.2.4.3 PLC and Inverter to Realize Multisection Speed Control of Three-phase Asynchronous Motor

Task Requirements

The output of the S7-1215C PLC is connected to the digital input terminal of the G120C PN inverter to realize multi-speed control of the motor. In this example, the first speed is set to 800r/min, the second speed is set to 400r/min, and the third speed is set to 1200r/min.

Finally, score the project after completion, as shown in Table 4-24.

Operation Steps

The operation steps are as follows:

(1) Wiring. The wiring work is completed between PLC and inverter, as shown in Figure 4-26.

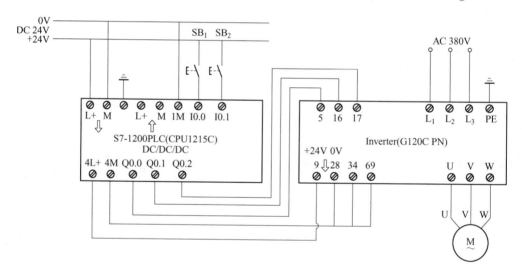

Figure 4-26 Three-stage speed control system wiring diagram

(2) Inverter parameter setting. The steps are as follows:

1) The basic commissioning of the inverter is required.

2) Select the motor parameters according to the actual motor parameters.

3) Select the predefined interface macro.

4) Motor speed related settings.

The inverter parameters is setted according to Table 4-22.

Table 4-22 Parameter settings

Par. No.	Parameter value	Features
P0015	1	Predefined macro parameter setting 1
P0840	r722.0	Using DI 0 as the start signal, r722.0 is the parameter of DI 0 state
P1000	3	Fixed frequency operation
P1016	1	Fixed speed mode adopts direct selection
P1022	722.4	Using DI 4 as a new signal for the selection of fixed settings, r722.4 is the parameter of DI 4 state
P1023	722.5	Using DI 5 as a new signal for the selection of fixed settings, r722.5 is the parameter of DI 5 state
P1003	800	Defining a fixed speed 3, unit is r/min
P1004	400	Defining a fixed speed 4, unit is r/min
P1070	1024	Define a fixed speed value as the main set value

(3) PLC. Hardware configuration refers to adding new equipment, selecting the corresponding PLC model of the training platform and setting the corresponding IP address. The operation steps are as follows:

1) Open the TIA Portal software and create an S7-1200 project.

2) Opening the project view, and clicking 'Add New Device', the Add New Device dialog box pops up, then selecting CPU1215 DC/DC/DC (6ES7 215-1AG40-0XB0).

3) Set the PLC's IP address to 192.168.0.1.

Software programming: Determine the I/O address allocation, fill in Table 4-23, and write the program according to the requirements.

Table 4-23 I/O address allocation table

	Input signal			Output signal		
Serial number	Function	Element	Address	Function	Inverter terminal name	Address

4) Write a program and download it to the PLC for debugging. The reference program is shown in Figure 4-27.

5) Connect the three-phase asynchronous motor to the line, and observe the running state after power-on.

6) Summarize and record the data.

Analysis and Summary

Summarize the task implementation process, find deficiencies, especially attach importance to the occurrence of malfunctions, and record Table 4-25 after analysis and summary.

4.2.5 Task Evaluation

Score the completion of the multisection speed control operation tasks of the three-phase asynchronous motor controlled by the PLC and the inverter, and fill in the score results in Table 4-24.

Network 1: .....

Comment

```
%I0.0      %I0.1      %M0.1                    %M0.0
"Tag_1"    "Tag_2"    "Tag_3"                   "Tag_4"
──┤ ├──────┤/├────────┤/├────────────────────────( )──

%M0.0                                             %DB3
"Tag_4"                                            "T1"
──┤ ├──                                           TON
                                                  Time                %M0.1
                                               IN      Q              "Tag_3"
                                      T# 20 S ─PT     ET──...          ─( )─
```

Network 2: .....

Comment

```
%I0.0                                                    %Q0.0
"Tag_1"                                                  "Tag_5"
──┤ ├───────────────────────────────────────────────────( S )──

                                                         %Q0.1
                                                         "Tag_6"
                                                        ─( S )──
```

Network 3: .....

Comment

```
%M0.1                                                    %Q0.1
"Tag_4"                                                  "Tag_6"
──┤ ├───────────────────────────────────────────────────( S )──
```

Network 4: .....

Comment

```
%I0.1                                                    %Q0.0
"Tag_2"                                                  "Tag_5"
──┤ ├───────────────────────────────────────────────( RESET_BF )──
                                                          3
```

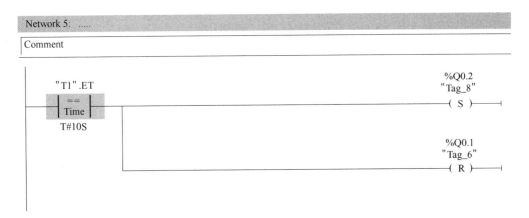

Figure 4-27 Reference program of three-section speed control system

Table 4-24 Score table

Task content	Assessment requirements	Grading	Distribution	Deduction	Score
Circuit design	Correctly design the wiring diagram of PLC and inverter to control the motor running circuit	(1) 5 points will be deducted for each missing function when the electrical control principle design function is incomplete; (2) If the wiring diagram is incorrectly expressed or the drawing is not standardized, 2 points will be deducted from each place	10		
Circuit wiring	Complete the wiring correctly according to the circuit diagram	(1) 5 points will be deducted for each wrong wire; (2) Points will be deducted in other cases as appropriate	10		
Inverter parameter setting and debugging	Set the corresponding parameters of the inverter according to the control requirements specified in the task	(1) 3 points will be deducted for each item when incompleting parameter setting of the inverter; (2) 2 points will be deducted for each parameter setting error; (3) The parameter will not be deducted by 20 points	20		

Continued Table 4-24

Task content	Assessment requirements	Grading	Distribution	Deduction	Score
PLC program writing and downloading	Input, download and monitor the PLC program according to the task requirements	(1) 15 points will be deducted when the software is unskilled and can not complete the input operation of the program; (2) 5 points will not be deducted for the IP address setting and downloading of the program; (3) 5 points will not be deducted for the operation of the Botu software monitoring program	20		
PLC control system operation demonstration	Demonstrate the start and stop of the motor correctly, and can explain it with the program and hardware	(1) The circuit can not realize the start deduction of 10 points; (2) 10 points will be deducted when the circuit cannot stop; (3) 10 points will be deducted for incorrect or incomplete functions; (4) 5 points will be deducted for each place where the operation phenomenon cannot be analyzed and explained	30		
Safe and civilized production	Abide by the 8S management system and the safety management system	(1) 10 points will be deducted for not wearing labor protection articles; (2) There are hidden safety hazards in the operation. 5 points will be deducted each time until the deduction is completed; (3) The operation site was not sorted out and rectified in time, 2 points will be deducted each time and the deduction is completed.	10		
		Total score			

4.2.6 Task Summary

The task implementation process record sheet is shown in Table 4-25.

Table 4-25 Task implementation process record sheet

Task implementation process record sheet		
Fault 1	Fault phenomenon	
	Cause of failure	
	Troubleshooting process	

Continued Table 4-25

<table>
<tr><td colspan="3">Task implementation process record sheet</td></tr>
<tr><td rowspan="3">Fault 2</td><td>Fault phenomenon</td><td></td></tr>
<tr><td>Cause of failure</td><td></td></tr>
<tr><td>Troubleshooting process</td><td></td></tr>
<tr><td rowspan="3">Fault 3</td><td>Fault phenomenon</td><td></td></tr>
<tr><td>Cause of failure</td><td></td></tr>
<tr><td>Troubleshooting process</td><td></td></tr>
<tr><td colspan="2">Summary</td><td></td></tr>
</table>

4.2.7 Task Development

Task requirements: Through the output of the S7-1215C PLC and the digital input terminal of the G120C inverter, the forward and reverse multisection control of the motor are realized. The three-section speed is set to 800r/min, 400r/min and 1200r/min. The specific control requirements are as follows:

(1) When the start button is pressed, the Q0.0 and Q0.1 outputs of the PLC are both 1, and the motor rotates forward and runs at a speed of 800r/min.

(2) After 5s, the Q0.0 and Q0.2 outputs of the PLC are both 1, and the motor rotates forward and runs at a speed of 400r/min.

(3) After 5s, the Q0.0 and Q0.2 outputs of the PLC are both 1, and the motor rotates forward and runs at a speed of 400r/min.

(4) After 10s, the Q0.0, Q0.1 and Q0.2 outputs of the PLC are all 1, and the motor rotates forward and runs at a speed of 1200r/min.

(5) After 15s, the Q0.3 and Q0.1 outputs of the PLC are both 1, and the motor reverses and runs at a speed of 800r/min.

(6) After 20s, the Q0.3 and Q0.2 outputs of the PLC are both 1, and the motor reverses and runs at a speed of 400r/min.

(7) After 25s, the Q0.3, Q0.1 and Q0.2 outputs of the PLC are all 1, and the motor reverses and runs at a speed of 1200r/min

(8) After 30s, the PLC output is 0 and the motor stops rotating.

(9) Whenever the emergency stop button is pressed, the motor stops immediately.

Selecting the predefined interface macro 1, the wiring diagram is shown in Figure 4-28.

4.2.8 Project summary

Through the implementation of Task 4.1, the basic structure composition and working principle of the inverter are studied. Taking SINAMICS G120C inverter as an example, on the basic of mastering the terminal function, and wiring Basic Operation Panel BOP-2 of the inverter, the basic operation of the inverter is mastered through the training of five operation task.

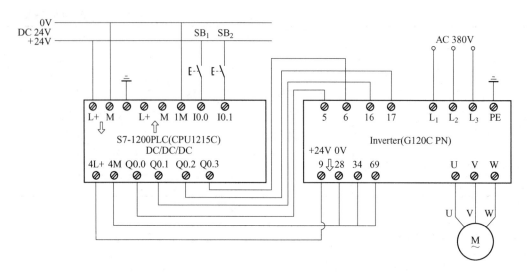

Figure 4-28 Wiring diagram of positive and negative three-section speed control system

Through the implementation of Task 4.2, the control of starting and stopping, forward and reverse rotation and multi-stage speed operation of three-phase asynchronous motor controlled by PLC and inverter are studied. Through the training of three operation tasks, the PLC control of Siemens G120 inverter wiring method is mastered; Using the predefined interface macro 1 of G120C PN converter, the function of the macro is preliminarily explored, which provides the foundation for the application of other macro functions; The basic operation of PLC is mastered, and three-phase asynchronous motor is inverter to control.

Exercises

(1) Briefly describe the structure and working principle of the inverter.

(2) How does the inverter change its direction during operation?

(3) How to realize the jog control of the motor? How to modify the jog frequency?

(4) What are the digital input and output terminals of G120C inverter?

(5) What are the analog input and output terminals of G120C inverter?

(6) What control methods does the inverter have?

(7) In addition to the pre-setting scheme 1, what other pre-setting schemes can realize the start and stop of three-phase asynchronous motors, forward and reverse rotation, and control of multisection speed operation?

(8) How to realize the frequency conversion speed regulation of analog quantity control?

(9) How to realize stepless speed regulation of inverter?

References

[1] Chen Baoling. Electrical control technology [M]. Dalian: Dalian University of Technology Press, 2019.

[2] Siemens. Botu V14 software user manual, 2017.

[3] Jin Zhe. Principle and application of PLC [M]. Beijing: Beijing Normal University Press, 2010.

[4] Chen Jianming. Electric Control and PLC Application: Based on S7-1200 PLC [M]. Beijing: Electronic Industry Press, 2020.

[5] Xiang Xiaohan. Siemens S7-1200 PLC learning manual: Programming based on LAD and SCL [M]. Beijing: Chemical Industry Press, 2018.

[6] Wu Fanhong. Siemens S7-1200 PLC application technology [M]. Beijing: Electronic Industry Press, 2017.

[7] Duan Licai et al. Siemens S7-1200 PLC Programming and User Guide [M]. Beijing: Machinery Industry Press, 2018.

[8] Khalid. Carmel, Ayman. Carmel. PLC industrial control [M]. Beijing: Machinery Industry Press, 2018.

[9] Ma hongqian, xu liange, shi jingbo, et al. PLC, frequency converter and touch screen technology and practice [M]. Beijing: electronics industry press, 2016.

[10] Shi Shouyong. S7-300 PLC, inverter and touch screen comprehensive application tutorial [M]. Beijing: Machinery Industry Press, 2018.

[11] Zhang Zhongquan. SINANICS G120 Inverter Control System User Manual. Beijing: Machinery Industry Press, 2016.

[12] Siemens. SINAMICS G120C Inverter Compact Operating Instructions, 2016.

项目1 电气控制与PLC

电气控制与PLC技术是装备制造大类专业关系密切的专业技术之一，在各种电机设备、数控机床和生产实践中广为应用。本项目分为电气原理图的识别与绘制、PLC基础知识和编程软件的学习三个任务。通过本项目的学习，使学生掌握电气原理图的基本知识，能够识别常用的低压电气元件，对PLC有一定了解，能够熟练的操作编程软件。这是本门课程的基础环节，为之后的具体项目任务学习打下坚实的基础。

任务1.1 电气控制与PLC基础知识

1.1.1 任务描述

本任务主要学习常用电气元件和电气原理图的基础知识，通过学习和观察低压电气元件，把相应的信息填入表格。绘制电气原理图，标出线号，完成简单的接线工作。最后总结任务实施的过程，并记录在相应表格中。

1.1.2 任务目标

（1）掌握常用低压电器元件的原理和用途。
（2）掌握电气原理图的基本绘制方法。
（3）掌握线号的标注方法。
（4）了解三相异步电动机的基本知识。
（5）培养学生安全操作及团队意识。

1.1.3 任务相关知识

1.1.3.1 低压断路器

低压断路器（曾称自动开关）是一种不仅可以接通和分断正常负荷电流和过负荷电流，还可以接通和分断短路电流的开关电器。低压断路器在电路中除起控制作用外，还具有一定的保护功能，如过负荷、短路、欠压和漏电保护等。低压断路器容量范围很大，最小为4A，而最大可达5000A。低压断路器广泛应用于低压配电系统，各种机械设备的电源控制和用电终端的控制和保护，其实物图及电气符号分别如图1-1和图1-2所示。

触头

与断路器主电路分、合机构机械上连动的触头，主要用于断路器分、合状态的显示，接在断路器的控制电路中通过断路器的分合，对其相关电器实施控制或连锁，比如向信号灯、继电器等输出信号。万能式断路器有六对触头（三常开、三常闭），DW45有八对触

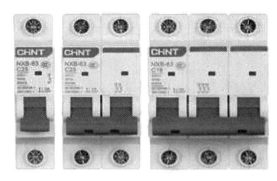

图 1-1 低压断路器实物图

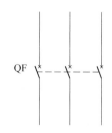

图 1-2 低压断路器电气符号

头（四常开、四常闭）。塑壳断路器额定电流 100A 为单断点转换触头，225A 及以上为桥式触头结构，约定发热电流为 3A；额定电流 400A 及以上可装两常开、两常闭，约定发热电流为 6A。操作性能次数与断路器的操作性能总次数相同。

用于断路器事故的报警触头，此触头只有当断路器脱扣分断后才动作。其主要用于断路器的负载出现过载短路或欠电压等故障时自由脱扣，报警触头从原来的常开位置转换成闭合位置，接通辅助线路中的指示灯或电铃、蜂鸣器等，显示或提醒断路器的故障脱扣状态。由于断路器发生因负载故障而自由脱扣的概率不太多，因而报警触头的寿命是断路器寿命的 1/10。报警触头的工作电流一般不会超过 1A。

脱扣器

脱扣器是一种用电压源激励的脱扣器，它的电压与主电路电压无关。分励脱扣器是一种远距离操纵分闸的附件。当电源电压等于额定控制电源电压的 70%~110% 时，就能可靠分断断路器。分励脱扣器是短时工作制，线圈通电时间一般不能超过 1s，否则线会被烧毁。塑壳断路器为防止线圈烧毁，在分励脱扣线圈串联一个微动开关，当分励脱扣器通过衔铁吸合，微动开关从常闭状态转换成常开。由于分励脱扣器电源的控制线路被切断，即使人为地按住按钮，分励线圈始终不再通电，从而避免了线圈烧损情况的产生。当断路器再扣合闸后，微动开关重新处于常闭位置。但万能式 DW45 产品在出厂时要由用户在使用时在分励脱扣器线圈之前串联一组常开触头。

欠电压脱扣器是在它的端电压降至某一规定范围时，使断路器有延时或无延时断开的一种脱扣器。当电源电压下降（甚至缓慢下降）到额定工作电压的 35%~70% 内，欠电压脱扣器应运作。欠电压脱扣器在电源电压等于脱扣器额定工作电压的 35% 时，欠电压脱扣器能防止断路器闭全；电源电压等于或大于 85% 欠电压脱扣器的额定工作电压时，在热态条件下，能保证断路器可靠闭合。因此，当受保护电路中电源电压发生一定的电压降时，能自动断开断路器切断电源，使该断路器以下的负载电器或电气设备免受欠电压的损坏。使用时，欠电压脱扣器线圈接在断路器电源侧，欠电压脱扣器通电后，断路器才能合闸，否则断路器合不上闸。

练一练

（1）观察低压断路器的外观，查看标签及技术参数。

（2）思考如何接线，把端子对应的线号和功能写在表 1-1 中。

表1-1 低压断路器端子功能表

序号	端子名称	功能	序号	端子名称	功能
1			5		
2			6		
3			7		
4			8		

1.1.3.2 熔断器

熔断器（Fuse）是指当电流超过规定值时，以本身产生的热量使熔体熔断，断开电路的一种电流保护器。熔断器是根据电流超过规定值一段时间后，以其自身产生的热量使熔体熔化，从而使电路断开。熔断器广泛应用于高低压配电系统和控制系统以及用电设备中，作为短路和过电流的保护器，是应用最普遍的保护器件之一。其实物图及电气符号分别如图1-3和图1-4所示。

图1-3 熔断器实物图

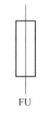

图1-4 熔断器电气符号

结构特性

熔体额定电流不等于熔断器额定电流，熔体额定电流按被保护设备的负荷电流选择，熔断器额定电流应大于熔体额定电流，与主电器配合确定。

熔断器主要由熔体、外壳和支座组成。其中熔体是控制熔断特性的关键元件，熔体的材料、尺寸和形状决定了熔断特性。熔体分为低熔点和高熔点两类，低熔点材料（如铅和铅合金）熔点低，容易熔断，由于其电阻率较大，故制成熔体的截面尺寸较大，熔断时产生的金属蒸汽较多，只适用于低分断能力的熔断器；高熔点材料（如铜、银）熔点高，不容易熔断，但由于其电阻率较低，可制成比低熔点熔体较小的截面尺寸，熔断时产生的金属蒸汽少，适用于高分断能力的熔断器。熔体的形状分为丝状和带状两种，改变截面的形状可显著改变熔断器的熔断特性。熔断器有各种不同的熔断特性曲线，可以适用于不同类型保护对象的需要。用于断路器事故的报警触头，只有当断路器脱扣分断后才动作，主要用于断路器的负载出现过载短路或欠电压等故障时而自由脱扣，报警触头从原来的常开位置转换成闭合位置，接通辅助线路中的指示灯或电铃、蜂鸣器等，显示或提醒断路器的故障脱扣状态。由于断路器发生因负载故障而自由脱扣的概率不多，因而报警触头的寿命是断路器寿命的1/10。报警触头的工作电流一般不会超过1A。

熔断器的动作是靠熔体的熔断来实现的，熔断器有个非常明显的特性，就是安秒特性。对熔体来说，其动作电流和动作时间特性即熔断器的安秒特性，也称作反时延特性。其特点是：过载电流小时，熔断时间长；过载电流大时，熔断时间短。

使用注意事项

熔断器的使用注意事项包括：

（1）熔断器的保护特性应与被保护对象的过载特性相适应，考虑到可能出现的短路电流，选用相应分断能力的熔断器。

（2）熔断器的额定电压要适应线路电压等级，熔断器的额定电流要大于或等于熔体额定电流。

（3）线路中各级熔断器熔体额定电流要相应配合，保持前一级熔体额定电流必须大于下一级熔体额定电流。

（4）熔断器的熔体要按要求使用相配合的熔体，不允许随意加大熔体或用其他导体代替熔体。

练一练

观察熔断器的外观，查看标签及技术参数。

1.1.3.3 交流接触器

接触器分为交流接触器（AC）和直流接触器（DC）应用于电力、配电与用电场合。接触器广义上是指工业中利用线圈流过电流产生磁场，使触头闭合，以达到控制负载的电器。其实物图及电气符号分别如图1-5和图1-6所示。

图1-5 交流接触器实物图

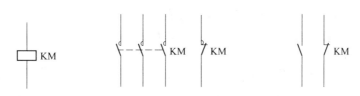

图1-6 交流接触器电气符号

接触器的工作原理是：当接触器线圈通电后，线圈电流会产生磁场，产生的磁场使静铁芯产生电磁吸力吸引动铁芯，并带动交流接触器点动作，常闭触点断开，常开触点闭合，两者是联动的。当线圈断电时，电磁吸力消失，衔铁在释放弹簧的作用下释放，使触点复原，常开触点断开，常闭触点闭合。直流接触器的工作原理跟温度开关的原理有点相似。

交流接触器利用主接点来控制电路，用辅助接点来导通控制回路。主接点一般是常开接点，而辅助接点常有两对常开接点和常闭接点，小型的接触器也经常作为中间继电器配合主电路使用。交流接触器的接点由银钨合金制成，具有良好的导电性和耐高温烧蚀性。交流接触器动作的动力源于交流通过带铁芯线圈产生的磁场，电磁铁芯由两个"山"字形的幼硅钢片叠成。其中一个是固定铁芯，套有线圈，工作电压可多种选择。为了使磁力稳定，铁芯的吸合面加上短路环，交流接触器在失电后，依靠弹簧复位。另一半是活动铁芯，构造和固定铁芯一样，用以带动主接点和辅助接点的闭合断开。20A以上的接触器加有灭弧罩，利用电路断开时产生的电磁力，快速拉断电弧，保护接点。

练一练

（1）观察交流接触器的外观，查看标签及技术参数。

（2）思考如何接线，把端子对应的线号和功能写在表1-2中。

表1-2 交流接触器端子功能表

序号	端子名称	功能	序号	端子名称	功能
1			8		
2			9		
3			10		
4			11		
5			12		
6			13		
7			14		

1.1.3.4 按钮

按钮是一种常用的控制电气元件，常用来接通或断开控制电路（其中电流很小），从而达到控制电动机或其他电气设备运行目的。按钮分为：

（1）常开按钮：开关触点断开的按钮；

（2）常闭按钮：开关触点接通的按钮；

（3）常开常闭按钮：开关触点既有接通也有断开的按钮；

（4）动作点击按钮：鼠标点击按钮，也称为按键，是一种电闸（或称开关），用来控制机械或程序的某些功能。

一般而言，红色按钮是用来使功能停止，而绿色按钮则可开始某一项功能。按钮的形状通常是圆形或方形，电子产品大都有用到按键这个最基本人机接口工具。随着工业水平的提升与创新，按键外观的也变得越来越多样化及丰富的视觉效果。其实物图及电气符号分别如图 1-7 和图 1-8 所示。

图 1-7 按钮实物图

图 1-8 按钮电气符号

练一练

(1) 观察按钮的外观，查看标签及技术参数。

(2) 思考如何接线，把端子对应的线号和功能写在表 1-3 中。

表 1-3 按钮端子功能表

序号	端子名称	功能	序号	端子名称	功能
1			4		
2			5		
3			6		

1.1.3.5 三相异步电动机

三相异步电动机是感应电动机的一种，是靠接入 380V 三相交流电流（相位差 120°）供电的一类电动机。由于三相异步电动机的转子与定子旋转磁场以相同的方向、不同的转速成旋转，存在转差率，所以叫三相异步电动机。三相异步电动机转子的转速低于旋转磁场的转速，转子绕组因与磁场间存在相对运动而产生电动势和电流，并与磁场相互作用产生电磁转矩，实现能量变换。

与单相异步电动机相比，三相异步电动机运行性能好，可节省各种材料。按转子结构的不同，三相异步电动机可分为笼式和绕线式两种。笼式转子的异步电动机结构简单、运行可靠、重量轻、价格便宜，得到了广泛的应用，但缺点是调速困难。绕线式三相异步电动机的转子和定子一样也设置了三相绕组并通过滑环、电刷与外部变阻器连接。调节变阻器电阻可以改善电动机的起动性能和调节电动机的转速。

三相异步电动机的结构

三相异步电动机是基于电磁原理把交流电能转换为机械能的一种旋转电机。三相鼠笼式异步电动机的基本结构有定子和转子两大部分。定子主要由定子铁芯、三相对称定子绕组和机座等组成,是电动机的静止部分。三相定子绕组一般有六根引出线,出线端装在机座外面的接线盒内,如图1-9所示。根据三相电源电压的不同,三相定子绕组可以接成星形（Y）或三角形（△）,如图1-10所示,然后与三相交流电源相连。转子主要由转子铁芯、转轴、鼠笼式转子绕组、风扇等组成,是电动机的旋转部分。小容量鼠笼式异步电动机的转子绕组大都采用铝浇铸而成,冷却方式一般都采用扇冷式。

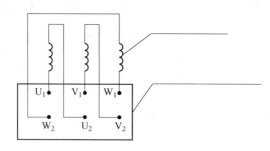

图1-9 三相定子绕组示意图

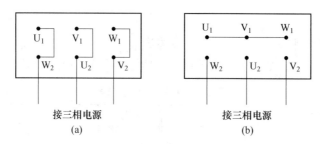

图1-10 三相定子绕组接法
（a）△型接法；(b) Y型接法

三相异步电动机的铭牌

三相异步电动机的额定值标记在电动机的铭牌上,三相鼠笼式异步电动机铭牌见表1-4。

表1-4 三相异步电动机的铭牌

型号	P_N/W	U_N/V	I_N/A
YS6324	180	380/660	0.65/0.38
$N_N/r \cdot min^{-1}$	接法	编号	绝缘等级
1400	Y/△	—	B

注：1. 功率：额定运行情况下,电动机轴上输出的机械功率。
  2. 电压：额定运行情况下,定子三相绕组应加电源的线电压值。
  3. 接法：定子三相绕组接法,当额定电压为380V/220V时,应为Y/△接法。
  4. 电流：额定运行情况下,当电动机输出额定功率时,定子电路的线电流大小。

三相异步电动机的检查

电动机使用前应做必要的检查，其包括：

（1）机械检查。机械检查主要检查引出线是否齐全、牢靠；转子转动是否灵活、匀称，有否异常声响等。

（2）电气检查。电动机在日常运行中常会有线圈松动，使绝缘磨损老化，或表面受污染、受潮等引起绝缘电阻日趋下降。绝缘电阻降低到一定值会影响电动机起动和正常运行，甚至会损坏电动机，危及人身安全。因此在各类电动机开始使用之前或经过霉季、受潮、重新安装之后，首先要测定各相绕组对机壳的绝缘电阻及绕组之间的绝缘电阻。绝缘电阻的测量一般用绝缘电阻表进行，学会绝缘电阻表的使用，在检查电机、电器及线路的绝缘情况和测量高值电阻时能给我们带来方便。

用绝缘电阻表检查电机绕组间及绕组与机壳之间的绝缘性能。用绝缘电阻表分别测试各相绕组始端 U_1、V_1、W_1 对机壳间以及各相绕组之间的绝缘电阻值。测量时将绝缘电阻表的接地端接至机壳（注意不要接触到涂漆之处，以免测量数据不准），另一测试端分别接到定子绕组的 U_1、V_1、W_1 端，然后以一定的速度（一般是 120rpm）摇转绝缘电阻表手柄，并保持手柄速度不变，读出绝缘电阻表读数。若此值大于 $0.5M\Omega$ 时，为合格；若小于 $0.5M\Omega$ 时，则说明该电机绝缘电阻下降，不经过维修或干燥处理电动机就不能再继续使用。若要测试两相绕组之间的绝缘电阻值，只需把绝缘电阻表的两测试端分别接到任意两相绕组的始端，用上述同样的方法摇转绝缘电阻表手柄，读出绝缘电阻表读数。由于绝缘电阻表在被摇转时，它的两个测试端间的电压可达 500V，故在测试时需注意安全。接线方法如图 1-11 所示。

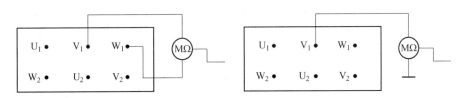

图 1-11　用绝缘电阻表检查电机绕组间及绕组与机壳之间的绝缘性能

练一练

（1）观察三相异步电动机的外观，查看标签及技术参数。

（2）思考如何接线，把端子对应的线号和功能写在表 1-5 中。

表 1-5　三相异步电动机端子功能表

序号	端子名称	功能	序号	端子名称	功能
1			5		
2			6		
3			7		
4			8		

1.1.4 任务实施

观察如图 1-12 所示的原理图，做到：

（1）说明其功能；

（2）绘制原理图；

（3）标出线号。

在实施过程中，如果出现故障，要对其进行故障的查找、分析和排除。注意安全意识和团队意识等职业核心素养的养成。最后对整个任务进行分析和总结，填入表 1-7 中。

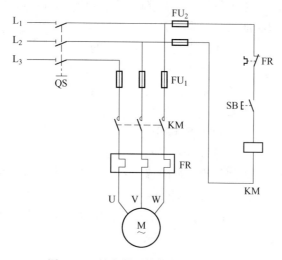

图 1-12　继电器系统控制的点动电路图

1.1.5 任务评价

任务评价表见表 1-6。

表 1-6　评分表

任务内容	考核要求	评分标准	配分	扣分	得分
准备工作	在任务实施之前做好充分的准备	准备工作是整个实施过程的入门阶段，这个阶段完成后才有资格进行下一步，如果这步未做或者失败，整个项目记 0 分	10		
工作原理描述	能够正确的说明整个系统的工作原理	（1）主电路功能描述 10 分； （2）控制回路功能描述 10 分	20		
原理图的绘制	正确完成原理图的绘制	（1）每接绘制错误一根线扣 5 分； （2）原理图工整 10 分； （3）原理图比例合适 10 分	30		

续表1-6

任务内容	考核要求	评分标准	配分	扣分	得分
线号的标注	正确完成原理图线号的标注	(1) 线号标注错误每处扣5分； (2) 线号漏标的每处扣5分； (3) 线号和实际元件不匹配的每处扣5分	30		
安全文明生产	遵守8S管理制度，遵守安全管理制度	(1) 未穿戴劳动保护用品扣10分； (2) 操作存在安全隐患，每次扣5分，扣完为止； (3) 操作现场未及时整理整顿，每次发现扣2分，扣完为止	10		
总　分					

1.1.6 任务小结

任务分析总结记录单见表1-7。

表1-7　任务分析总结记录单

任务分析总结记录单		
故障1	故障过程	
	故障原因	
	排除过程	
故障2	故障现象	
	故障原因	
	排除过程	
故障3	故障现象	
	故障原因	
	排除过程	
总　结		

1.1.7 任务拓展

任务要求：对应图1-12，完成线路的连接，并分析整个过程。

任务1.2　PLC基础知识

1.2.1 任务描述

可编程控制器（PLC）是种专门为在工业环境下应用而设计的数字运算操作电子系统。它采用一种可编程的存储器，在其内部存储执行逻辑运算、顺序控制、定时、计数和算术运算等操作的指令，通过数字式或模拟式的输入输出来控制各种类型的机械设备或生产过程。通过本任务的学习，对 PLC 有初步的了解，包括产生、定义、特点发展等，为后续 PLC 编程打下良好的基础。

1.2.2 任务目标

(1) 了解 PLC 的产生及定义。
(2) 了解 PLC 的特点。
(3) 了解 PLC 的发展方向。
(4) 了解 1200 系列 PLC。
(5) 掌握 PLC 输入端的接线方式。
(6) 培养学生安全操作及团队意识。

1.2.3 任务相关知识

本课程使用的是西门子 1215C 系列 PLC。可编程序控制器（Programmable Controller）是一种通用工业控制计算机，是以微处理器为基础，运用计算机技术、微电子技术、自动控制技术、数字技术和网络通信技术而发展起来。它面向过程、面向用户，适应工业环境，操作方便，可靠性高，已成为现代工业控制的四大支柱（PLC 技术、机器人技术、CAD/CAM、数控技术）之一。它的控制技术代表着当前程序控制的先进水平，并已经成为自动控制系统的基本装置。

最初的可编程序逻辑控制器（Programmable Logic Controller）以逻辑控制为主，故简称为 PLC。现在，可编程序控制器的功能在不断扩展，除了逻辑控制外，还增加了模拟量调节、数值运算、监控、通信联网等功能，故将其改称为可编程序控制器，简称 PC。但为了与个人电脑 PC（Personal Computer）相区别，还有许多人将其简称为 PLC。

1.2.3.1 PLC 的产生及定义

20 世纪 60 年代末，工业生产大多以大批量、少品种生产方式为主，而这种大规模生产线的控制以继电器控制系统占主导地位。由于市场的发展，要求工业生产发展方向小批量、多品种生产方式转变，这样继电器控制系统就需要重新设计安装，十分费时、费工、费料，阻碍了更新周期的缩短。为了改变这种状况，1968 年美国通用汽车公司（GM）对外公开招标，期望设计出一种新型的自动工业控制装置，来取代继电器控制装置，从而达到汽车型号不断更新的目的。为此提出了以下 10 项指标：

(1) 编程方便，现场可修改程序。
(2) 维修方便，采用插件式结构。
(3) 可靠性高于继电器控制装置。
(4) 可将数据直接送入管理计算机。
(5) 输入可以是交流 115V。
(6) 输出为交流 115V、2A 以上，能直接驱动电磁阀和接触器等。
(7) 用户存储容量至少可以扩展到 4kB。
(8) 体积小于继电器控制装置。

(9)扩展时原系统变更较小。

(10)成本可与继电器控制装置竞争。

1969年美国数字设备公司（DEC）根据招标要求研制出了世界第一台可编程序逻辑控制器（PLC），并应用于美国通用汽车公司自动装配线上，获得成功。从此PLC在美国其他工业领域广泛应用，开创了工业控制的新时代。

国际电工委员会（IEC）曾先后3次颁布了可编程序控制器标准草案，1987年2月第3次草案对可编程序控制器的定义是："可编程序控制器是一种数字运算操作的电子系统，专为在工业环境下应用而设计，它采用了可编程序的存储器，用来在其内部存储执行逻辑运算、顺序控制、定时、计数和算术运算等操作的指令，并通过数字式和模拟式的输入和输出，控制各种类型机械的生产过程。可编程序控制器及其有关外围设备都应按易于与工业系统联成一个整体，易于扩充其功能的原则设计"。由上述定义可看出，可编程序控制器PLC首先是一台计算机，而且是专为工业环境下应用而设计的工业计算机。由于是根据使用者提出的指标而设计产生的，同时又具有其他工业控制设备很难兼具的独一无二的特性，其广泛适应于工业上的各种控制。

1.2.3.2 PLC的特点

可靠性高、抗干扰能力强

PLC是专为工业控制环境而设计的，因此可靠性高、抗干扰性强是其最主要的特点之一。由于在设计中采取了一系列措施，所以PLC的平均无故障工作时间达几十万小时。可以这样说，到目前为止，还没有任何一种工业控制设备的可靠性可以达到PLC水平，而且随着器件技术的提高，PLC的可靠性还将会继续提高。

一般的数字电子设备产生故障的原因可分为两类：一类是由恶劣的外界环境，如电磁干扰、高温、高湿、振动、有害气体而引起的故障；另一类是由内部器件的老化、失效、存储器信息的丢失、错误程序的运行等而引起的故障。为此，可从硬件和软件两方面采取措施减少故障的产生。

硬件方面采取模块式结构，并采取光电隔离、屏蔽、滤波等抗干扰措施。另外还对某些模块设置输出连锁保护、环境检测、自诊断电路等。

软件方面设置自诊断、实时监控、信息保护等程序。另外，大型PLC还采取了双CPU冗余系统或CPU表决系统，来进一步提高PLC的可靠性。

编程简单、易于使用

这是PLC又一个主要的特点。PLC大多采用与继电器控制电路类似的梯形图进行编程，直观性强，一般工程技术人员很容易掌握，并且易于操作和使用。

PLC还设计了其他种类的编程语言，如指令语句表编程语言、功能块编程语言等，以适应不同编程人员的需要，更好地完成各种控制功能。

功能强、通用性好、使用灵活

现代PLC运用计算机技术、微电子技术、数字技术、网络通信技术和集成工艺等最新技术，增强了其复杂控制及通信联网等功能。目前的PLC产品已经实现系列化、模块化、标准化，可灵活方便地组成不同规模、不同功能的控制系统，来满足用户的需求。

便于安装调试

由于 PLC 功能齐全，只要能够合理选择各种模块组成系统，就无须另外配置硬件，同时也无须软件的二次开发。PLC 的应用程序也可方便地在实验室模拟调试，调试成功后，再到现场安装调试。

体积小、重量轻、功耗低

由于 PLC 采用了微电子技术，因而体积小、结构紧凑、重量轻、功耗低。

1.2.3.3 PLC 的发展方向

随着计算机技术、数字技术、半导体集成技术、网络通信技术等高新技术的发展，PLC 也得到了飞速的发展。目前，PLC 已广泛地应用于各个领域。

PLC 的发展方向大体分为两个：一是向体积更小、功能更强、价格更低的小型化方向发展，提供性能价格比更高的小型 PLC 控制系统，使之应用范围更加广泛；二是向速度更快、功能更多、联网与通信能力更强的大型化方向发展，提供高性能、高速度、高性价比的大、中型 PLC 控制系统，以适应大规模、复杂控制系统的需要。具体体现在以下几个方面：

（1）网络通信功能增强。
（2）发展智能输入/输出模块。
（3）采用多样化编程语言。
（4）增强外部故障检测及处理能力。

1.2.3.4 1200 系列 PLC 简介

SIMATIC S7-1200 是一款紧凑型、模块化的 PLC，可完成简单逻辑控制、高级逻辑控制、HMI 和网络通信等任务。其对于需要网络通信功能和单屏或多屏 HMI 的自动化系统，易于设计和实施，并具有支持小型运动控制系统、过程控制系统的高级应用功能。新的模块化 SIMATIC S7-1200 控制器是西门子新推出产品的核心，可实现简单却高度精确的自动化任务。SIMATIC S7-1200 控制器实现了模块化和紧凑型设计，功能强大、投资安全并且完全适合各种应用。其可扩展性强、灵活度高的设计，其可实现最高标准工业通信的通信接口以及一整套强大的集成技术功能，使该控制器成为完整、全面的自动化解决方案的重要组成部分。其外形如图 1-13 所示。

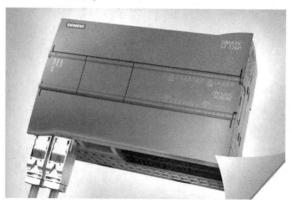

图 1-13　西门子 1200 系列 PLC 外形

优势

SIMATIC HMI 基础面板的性能经过优化，旨在与这个新控制器以及强大的集成工程组态完美兼容，可确保实现简化开发、快速启动、精确监控和最高等级的可用性。正是这些产品之间的相互协同及其创新性的功能，可将小型自动化系统的效率提升到一个前所未有的水平。

该面板用于可扩展设计中紧凑自动化的模块化概念。实现了通信简便，并满足一系列的独立自动化系统的应用需求，在工程组态中实现最高效率。使用完全集成的新工程组态 SIMATIC STEP 7 Basic，并借助 SIMATIC WinCC Basic 对 SIMATIC S7-1200 进行编程。SIMATIC STEP 7 Basic 的设计理念是直观、易学和易用。这种设计理念可以在工程组态中实现最高效率。一些智能功能，例如直观编辑器、拖放功能和"IntelliSense"（智能感知）工具，能让工程进行的更加迅速。这款新软件的体系结构源于对未来创新的不断追求，西门子在软件开发领域已经有很多年的经验，因此 SIMATIC STEP 7 的设计是以未来为导向的。

设计和功能

SIMATIC S7-1200 系统有五种不同模块，分别为 CPU 1211C、CPU 1212C、CPU 1214C、CPU 1215C 和 CPU 1217C。其中每一种模块都可以进行扩展，以满足系统需要。可在任何 CPU 的前方加入一个信号板，轻松扩展数字或模拟量 I/O，同时不影响控制器的实际大小。可将信号模块连接至 CPU 的右侧，进一步扩展数字量或模拟量 I/O 容量。CPU 1212C 可连接两个信号模块，CPU 1214C、CPU 1215C 和 CPU 1217C 可连接 8 个信号模块。最后，所有的 SIMATIC S7-1200 CPU 控制器的左侧均可连接多达 3 个通信模块，便于实现端到端的串行通信。

所有的 SIMATIC S7-1200 硬件都有内置的卡扣，可简单方便地安装在标准的 35mm DIN 导轨上。这些内置的卡扣也可以卡入已扩展的位置，当需要安装面板时，可提供安装孔。SIMATIC S7-1200 硬件可以安装在水平或竖直的位置，提供其他安装选项。这些集成的功能在安装过程中为用户提供了最大的灵活性，并使 SIMATIC S7-1200 为各种应用提供了实用的解决方案。

所有的 SIMATIC S7-1200 硬件都经过专门设计，以节省控制面板的空间。例如，经过测量，CPU 1214C 的宽度仅为 110mm，CPU 1212C 和 CPU 1211C 的宽度仅为 90mm。结合通信模块和信号模块的较小占用空间，在安装过程中，该模块化的紧凑系统节省了宝贵的空间，并提供了最高效率和最大灵活性。SIMATIC S7-1200 可扩展的紧凑自动化的模块化概念实现了简便的通信，并能完全满足一系列的独立自动化需求。

1.2.3.5　PLC 电源及输入端的接线

CPU 1215C DC/DC/DC（6ES7 215-1AG40-0XB0）的外部接线如图 1-14 所示。由图可知，PLC 供电电源为直流 24V，其中 L+接电源正极，M 接电源负极。1M 输入的公共端与输入端子、输入开关和直流 24V 电源连接后形成一个回路；2M 为模拟量输出公共端；3M 为模拟量输入公共端；4M 为 PLC 输出公共端。

练一练

（1）观察 XK-SX2C 高级维修电工实训台上面的 PLC，查看标签及技术参数。

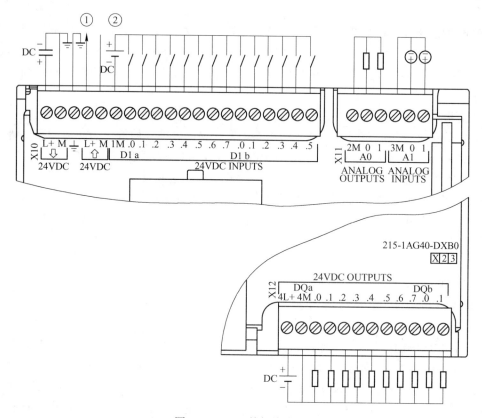

图 1-14 PLC 外部接线图

(2) 思考如何接线，把端子对应的线号和功能写在表 1-8 中。

表 1-8 PLC 端子功能表

序号	端子名称	功能	序号	端子名称	功能
1			14		
2			15		
3			16		
4			17		
5			18		
6			19		
7			20		
8			21		
9			22		
10			23		
11			24		
12			25		
13			26		

续表 1-8

序号	端子名称	功能	序号	端子名称	功能
27			36		
28			37		
29			38		
30			39		
31			40		
32			41		
33			42		
34			43		
35			44		

1.2.4 任务实施

在 XK-SX2C 高级维修电工实训台上，使用跨接线完成 PLC 与电源之间的连接，在确保连接正确的基础上，接通电源。然后接入两个开关，在确保连接正确的基础上，接通电源，按下开关，观察 PLC 上面指示灯的运行情况。

1.2.4.1 任务要求

参照图 1-14，对 PLC 进行通电测试，接入两个按钮进行测试。完成后对本项目进行评分，详见表 1-9。

1.2.4.2 操作步骤

其操作步骤分别为：
（1）对所需元件进行故障查找，确保每个元件能正常使用。
（2）绘制原理图，并标出线号。
（3）连接 PLC 的电源线并用万用表测试。
（4）连接两个开关按钮。
（5）用万用表做通电前的检查工作。
（6）通电观察系统的运行状态。
（7）总结并记录。

1.2.4.3 分析总结

总结整个任务实施过程，查找不足，尤其对于出现的故障要高度重视，分析总结后记录在表 1-10 中。

1.2.5 任务评价

评分表见表 1-9。

表 1-9 评分表

任务内容	考核要求	评分标准	配分	扣分	得分
准备工作	(1) 对所需元件进行故障查找，确保每个元件能正常使用； (2) 绘制原理图	准备工作是整个实施过程的入门阶段，这个阶段完成后才有资格进行下一步，如果这步未做或者失败，整个项目记 0 分	10		
电源线的连接	能够正确的给 PLC 供电	(1) 正极连接不正确扣 5 分； (2) 负极连接不正确扣 5 分	10		
输入开关按钮的连接	正确的连接两个输入开关按钮	(1) 每接错一个扣 10 分； (2) 不能形成回路扣 10 分； (3) 常开、常闭触点连接错误每处扣 5 分	30		
通电前的安全测试	给 PLC 通电前要进行安全测试，确保没问题后才能通电	确保没有短路、断路、虚接等现象，此步骤不做整个项目记 0 分	10		
通电运行演示	正确演示输入开关按钮，并能结合硬件进行说明	不能对操作现象进行分析和说明的每处扣 5 分	30		
安全文明生产	遵守 8S 管理制度，遵守安全管理制度	(1) 未穿戴劳动保护用品，扣除 10 分； (2) 操作存在安全隐患，每次扣除 5 分，扣完为止； (3) 操作现场未及时整理整顿，每次发现扣除 2 分，扣完为止	10		
总 分					

1.2.6 任务小结

任务分析总结记录单见表 1-10。

表 1-10 任务分析总结记录单

任务分析总结记录单			
故障 1	故障现象		
	故障原因		
	排除过程		
故障 2	故障现象		
	故障原因		
	排除过程		

续表 1-10

		任务分析总结记录单
故障3	故障现象	
	故障原因	
	排除过程	
	总　结	

1.2.7 任务拓展

任务要求：学习 8S 管理制度，在实际的学习和工作中使用。8S 包括整理（SEIRI）、整顿（SEITON）、清扫（SEISO）、清洁（SETKETSU）、素养（SHITSUKE）、安全（SAFETY）、节约（SAVE）、学习（STUDY）八个项目，因其古罗马发音均以"S"开头，简称为 8S。8S 管理法的目的，是使企业在现场管理的基础上，通过创建学习型组织不断提升企业文化的素养，消除安全隐患、节约成本和时间。使企业在激烈的竞争中，永远立于不败之地。

1.2.7.1 整理

整理是指将混乱的状态收拾成井然有序的状态。

其目的包括：

(1) 腾出空间，空间活用，增加作业面积。

(2) 物流畅通、防止误用、误送等。

(3) 塑造清爽的工作场所。

实施要领包括：

(1) 自己的工作场所（范围）全面检查，包括看得到和看不到的。

(2) 制定"要"和"不要"的判别基准。

(3) 将不要物品清除出工作场所，要有决心。

(4) 对需要的物品调查使用频度，决定日常用量及放置位置。

(5) 制订废弃物处理方法。

(6) 每日自我检查。

1.2.7.2 整顿

整理是指通过前一步整理后，对生产现场需要留下的物品进行科学合理的布置和摆放，以便用最快的速度取得所需之物，在最有效的规章、制度和最简捷的流程下完成作业。

其目的包括：

(1) 使工作场所一目了然，创造整整齐齐的工作环境。

(2) 不用浪费时间找东西，能在 30s 内找到要找的东西，并能立即使用。

实施要领包括：

(1) 前一步骤整理的工作要落实。

(2) 流程布置，确定放置场所、明确数量，即：
1) 物品的放置场所原则上要100%设定；
2) 物品的保管要定点（放在哪里合适）、定容（用什么容器、颜色）、定量（规定合适数量）；
3) 生产线附近只能放真正需要的物品。
(3) 规定放置方法，即：
1) 易取，提高效率；
2) 不超出所规定的范围；
3) 在放置方法上多下功夫。
(4) 划线定位。
(5) 场所、物品标识须被满足。

整改标识如下所示：
(1) 放置场所和物品标识原则上一一对应；
(2) 标识方法全公司要统一。

1.2.7.3 清扫

清扫是指清除工作场所内的脏污，并防止污染的发生，将岗位保持在无垃圾、无灰尘、干净整洁的状态。清扫的对象包括地板、墙壁、工作台、工具架、工具柜等，机器、工具、测量用具等。

其目的包括：
(1) 消除脏污，保持工作场所干净、亮丽的环境，使员工保持一个良好的工作情绪。
(2) 稳定品质，最终达到企业生产零故障和零损耗。

实施要领包括：
(1) 建立清扫责任区（工作区内外）。
(2) 执行例行扫除，清理脏污，形成责任与制度。
(3) 调查污染源，予以杜绝或隔离。
(4) 建立清扫基准，作为规范。

1.2.7.4 清洁

清洁是指将上面的3S（整理、整顿、清扫）实施的做法进行到底，形成制度，并贯彻执行及维持结果。

其目的是为了维持上面3S的成果，并显现"异常"之所在。

实施要领包括：
(1) 前面3S工作实施彻底。
(2) 定期检查，实行奖惩制度，加强执行。
(3) 管理人员经常带头巡查，以表重视。

1.2.7.5 素养

素养是指人人依规定行事，从心态上养成能随时进行8S管理的好习惯并坚持下去。

其目的包括：

（1）提高员工素质，培养员工成为一个遵守规章制度，并具有良好工作素养习惯的人。

（2）营造团体精神。

实施要领包括：

（1）培训共同遵守的有关规则和规定。

（2）新进人员强化教育和实践。

1.2.7.6 安全

安全是指清除安全隐患，保证工作现场员工人身安全及产品质量安全，预防意外事故的发生。

其目的包括：

（1）规范操作、确保产品质量、杜绝安全事故。

（2）保障员工的人身安全，保证生产连续安全正常的进行。

（3）减少因安全事故而带来的经济损失。

实施要领包括：

（1）制定正确作业流程，适时监督指导。

（2）对不合安全规定的因素及时发现消除，所有设备都进行清洁、检修，能预先发现存在的问题，从而消除安全隐患。

（3）在作业现场彻底推行安全活动，使员工对于安全用电、确保通道畅通、遵守搬用物品的要点养成习惯，建立有规律的作业现场。

（4）员工正确使用保护器具，不违规作业。

1.2.7.7 节约

节约是指对时间、空间、资源等方面合理利用，减少浪费，降低成本，以发挥它们的最大效能，从而创造一个高效率的，物尽其用的工作场所。

其目的是为了养成降低成本习惯，培养作业人员减少浪费的意识。

实施要领包括：

（1）以自己就是主人的心态对待企业的资源。

（2）能用的东西尽可能利用。

（3）切勿随意丢弃，丢弃前要思考其剩余之使用价值。

（4）减少动作浪费，提高作业效率。

（5）加强时间管理意识。

1.2.7.8 学习

学习是指深入学习各项专业技术知识，从实践和书本中获取知识，同时不断地向同事及上级主管学习。

其目的包括：

（1）学习长处，完善自我，提升自己综合素质。

（2）让员工能更好的发展，从而带动企业产生新的动力去应对未来可能存在的竞争与变化。

实施要领包括：

（1）学习各种新的技能技巧，才能不断去的满足个人及公司发展的需求。

（2）与人共享，能达到互补、互利，制造共赢，互补知识面与技术面的薄弱，互补能力的缺陷，提升整体的竞争力与应变能力。

（3）内部外部客户服务的意识，为集体（或个人）的利益或为事业工作，服务有关的同事和客户（如注意内部客户的服务）。

任务1.3　编程软件的使用

1.3.1　任务描述

博途（TIA）是全集成自动化软件 TIA portal 的简称，英语全称是 Totally Integrated Automation Portal，是西门子工业自动化集团发布的一款全新的全集成自动化软件。它是业内首个采用统一的工程组态和软件项目环境的自动化软件，几乎适用于所有自动化任务。借助该全新的工程技术软件平台，用户能够快速、直观地开发和调试自动化系统。博途作为一切未来软件工程组态包的基础，可对西门子全集成自动化中所涉及的所有自动化和驱动产品进行组态、编程和调试。

通过本任务的学习，掌握在博途编程软件中完成程序的输入、下载和监控的能力。把任务实施过程中遇到的问题进行总结，并记录在相应表格中。

1.3.2　任务目标

（1）掌握软件的输入方法。

（2）掌握硬件组态设备的配置方法。

（3）掌握软件监控的方法。

（4）掌握查找、分析和排除故障的能力。

（5）培养学生安全操作及团队意识。

1.3.3　任务相关知识

博途（TIA）编程软件的使用方法和步骤如下：

（1）双击博途软件图标，在出现的博途软件界面中，单击识图左面的"创建新项目"，可以对新项目进行名称、路径等进行更改后，单击"创建"按钮，如图1-15所示。

（2）在弹出的窗口中，单击"组态设备"图标，进行硬件设备组态，如图1-16所示。

（3）在出现的窗口中，单击"控制器"图标，在出现的控制器列表中，选择"CPU 1215C DC/DC/DC"，并按照 PLC 实际的版本号进行选择，然后单击按钮"添加"，如图1-17所示。

（4）在出现的窗口中，用鼠标单击选中 PLC 的以太网口，在窗口下半部分，单击"属性"标签，在左面单击"以太网地址"，进行 PLC 的 IP 地址设置，如图1-18所示。

图 1-15 创建新项目

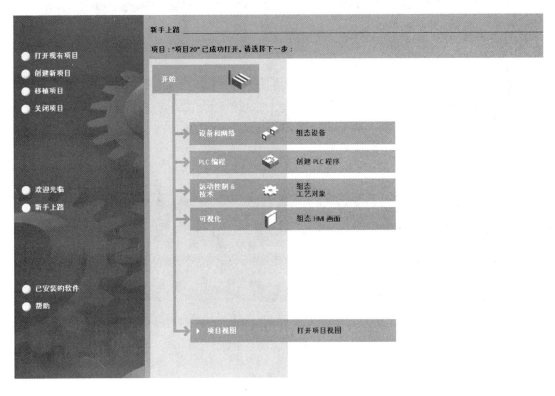

图 1-16 硬件组态

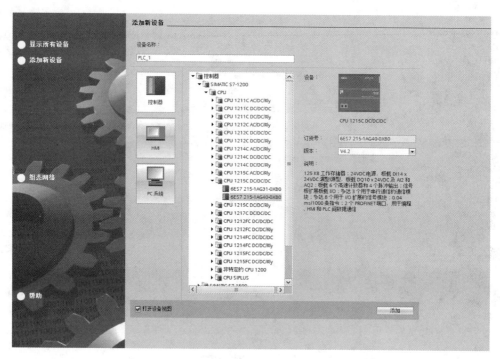

图 1-17　PLC 硬件添加

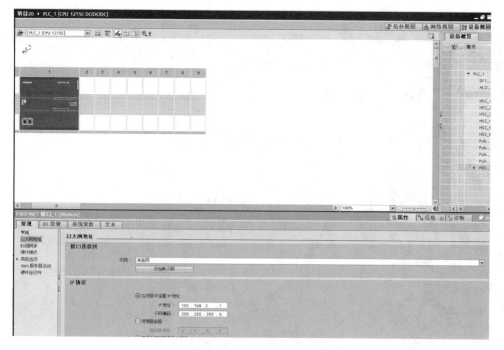

图 1-18　PLC 地址设置

（5）选中 PLC 后，单击 图标，将硬件组态下载到 PLC 中，如图 1-19 所示。

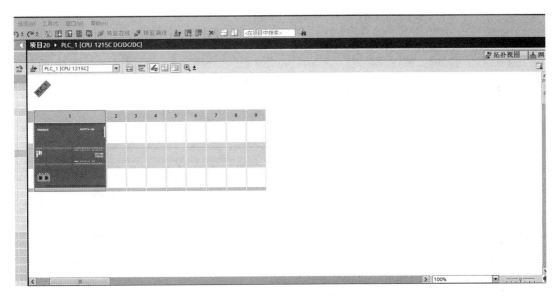

图 1-19 硬件组态下载

(6) 在弹出的窗口中,单击"开始搜索"按钮,寻找目标 PLC 及其 IP 地址,如图 1-20所示。

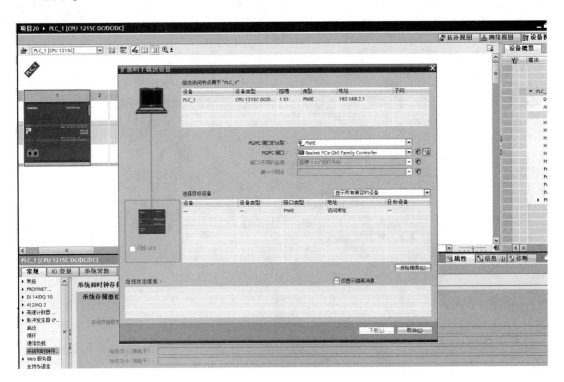

图 1-20 寻找目标 PLC

(7) 找到并连接到目标 PLC 后,单击"下载"按钮,将硬件组态下载到 PLC 中,如

图 1-21 所示。在下载过程中若出现窗口如图 1-22 所示内容,单击"在不同情况下继续",继续完成下载。

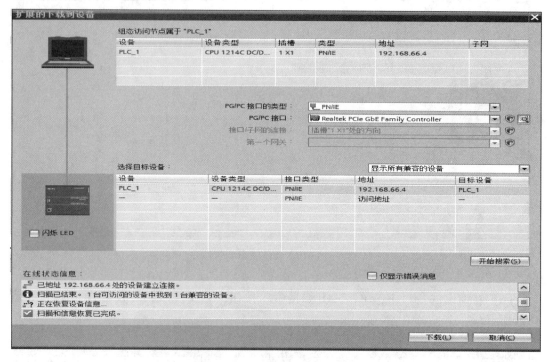

图 1-21 连接 PLC 并下载 (1)

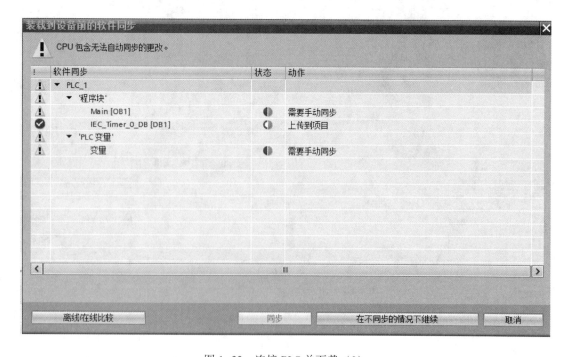

图 1-22 连接 PLC 并下载 (2)

（8）在随后出现的窗口中的"停止模块"选项，选择"全部停止"后，鼠标单击"装载"按钮，将 PLC 硬件下载到 PLC 中，如图 1-23 所示。

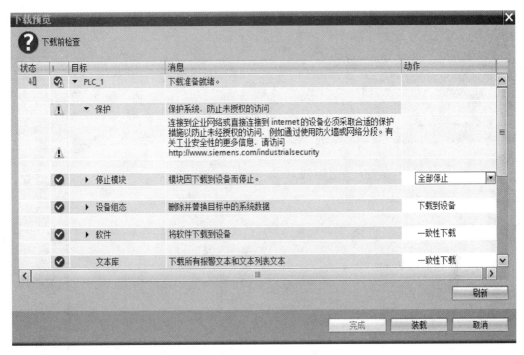

图 1-23　下载前检查

（9）硬件下载完成后，展开"设备"选项卡下的"程序块"，双击"Main(OB1)"，打开编程界面，如图 1-24 所示。

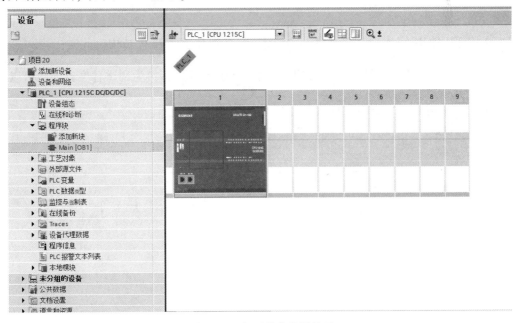

图 1-24　打开程序编辑界面

（10）在打开的编程界面，可以通过拖拽或双击如图 1-25 所示区域 1 或区域 2 的相应的指令，将其添加到逻辑行上，对于常用的指令，可以将图 1-25 中区域 1 的指令拖拽到区域 2，以增加编程的便捷性。可通过图标 增加逻辑行，也可以通过图标 删除逻辑行。

图 1-25　编程界面介绍

（11）程序编辑完成后，通过单击 图标，将程序下载到 PLC 中。此时，若 PLC 显示为停止状态，可以单击图标 ，将 PLC 启动处于运行状态，为调试程序方便，可以单击 ，进行程序监控，查看程序的运行情况，如图 1-26 所示。

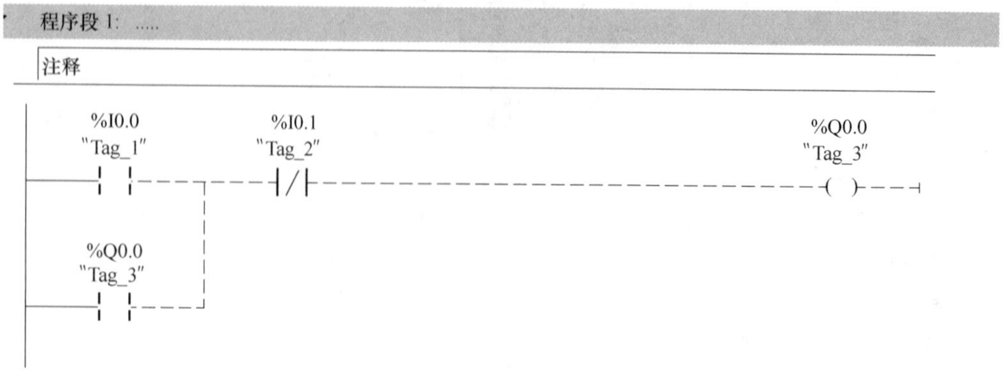

图 1-26　程序运行监控

练一练

输入图1-26的程序,下载到PLC,并打开监控观察运行状态。

1.3.4 任务实施

1.3.4.1 任务要求

能熟练掌握编程软件。在XK-SX2C高级维修电工实训台上,按照如图1-27所示中的输入点进行PLC的接线,并下载程序。在确保连接正确的基础上,接通电源,对输入按钮开关进行操作,观察PLC的运行状态。在实施过程中,如果出现故障,要对其进行故障的查找、分析和排除。注意安全意识和团队意识等职业核心素养的养成。

完成后对本项目进行评分,详见表1-12。

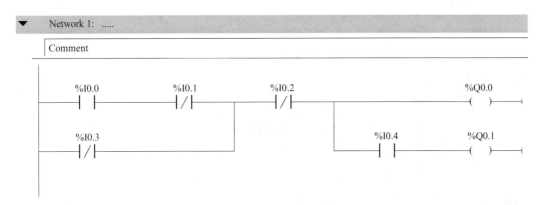

图1-27 参考程序

1.3.4.2 操作步骤

操作步骤分别为:

(1) 对所需元件进行故障查找,确保每个元件能正常使用。
(2) 绘制原理图。
(3) 连接PLC的电源线,用万用表测量,确保无误后方可通电。
(4) 确定I/O地址分配,并填入表1-11中。

表1-11 I/O地址分配表

输入信号				输出信号			
序号	功能	元件	地址	序号	控制对象	元件	地址
1				1			
2				2			
3				3			
4				4			
5				5			

（5）根据 I/O 地址分配完成输入端的线路连接。

（6）检查无误后再给 PLC 通电。

（7）把图 1-27 中的程序下载到 PLC 中进行调试。

（8）用万用表做通电前的检查工作，若出现故障及时排除并记录总结。

（9）通电观察系统的运行状态，若出现故障及时排除并记录总结。

（10）总结并记录。

1.3.4.3 分析总结

总结整个任务实施过程，查找不足，尤其对于出现的故障要高度重视，分析总结后记录在表 1-13 中。

1.3.5 任务评价

评分表见表 1-12。

表 1-12 评分表

任务内容	考核要求	评分标准	配分	扣分	得分
准备工作	（1）对所需元件进行故障查找，确保每个元件能正常使用； （2）绘制原理图	准备工作是整个实施过程的入门阶段，这个阶段完成后才有资格进行下一步，如果这步未做或者失败，整个项目记 0 分	10		
工作原理描述	能够正确的说明梯形图的工作原理	每说错一个输出的功能扣 10 分	10		
PLC 系统的接线	正确完成硬件接线	（1）每接错一根线扣 5 分； （2）其他情况酌情扣分	20		
PLC 程序输入与下载	按照任务要求将 PLC 程序输入、下载与监控	（1）软件使用不熟练，不能完成程序的输入操作扣除 15 分； （2）不会进行程序的 IP 地址设置与下载扣 5 分； （3）不会操作博途软件监控程序的运行扣 5 分	20		
PLC 控制系统运行演示	正确程序，并能结合程序、硬件进行说明	（1）不能实现输出的启动，每处扣 10 分； （2）不能对操作现象进行分析和说明的每处扣 5 分	30		
安全文明生产	遵守 8S 管理制度，遵守安全管理制度	（1）未穿戴劳动保护用品，扣除 10 分； （2）操作存在安全隐患，每次扣除 5 分，扣完为止； （3）操作现场未及时整理整顿，每次发现扣除 2 分，扣完为止	10		
总　　分					

1.3.6 任务小结

任务分析总结记录单见表1-13。

表1-13 任务分析总结记录单

任务分析总结记录单		
故障1	故障现象	
	故障原因	
	排除过程	
故障2	故障现象	
	故障原因	
	排除过程	
故障3	故障现象	
	故障原因	
	排除过程	
总　结		

1.3.7 任务拓展

任务要求：在XK-SX2C高级维修电工实训台上，按照如图1-28中的输入点进行PLC的接线，并下载程序。在确保连接正确的基础上，接通电源，对输入按钮开关进行操作，观察PLC的运行状态。

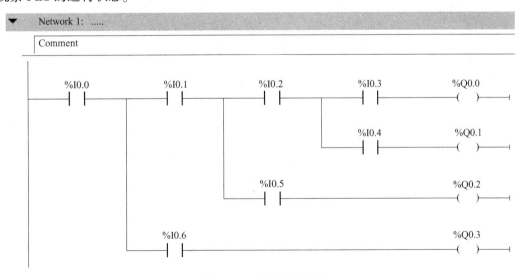

图1-28 任务拓展程序

1.3.8 项目小结

本项目包含 3 个任务，是深入学习电气控制与 PLC 的基础，这个知识技能一定要打牢固。任务 1.1 主要介绍了常用的低压断路器、熔断器、接触器、开关按钮的作用、工作原理及图形符号；任务 1.2 介绍了 PLC 的产生、定义、特点及发展，还重点讲述了 1200 系列 PLC 的相关内容，这部分属于识记的内容；任务 1.3 介绍了软件的用法和程序的测试方法，注意硬件组态的构建以及 IP 地址的匹配。PLC 的外部接线图也是需要重点掌握的内容。

练习题

（1）简述 PLC 的发展历程。
（2）简述博途（TIA）软件的优点。
（3）查阅资料，分析比较 1200 系列 PLC 的参数。
（4）简述西门子 1215C 的外部接线方式。
（5）输入下面程序（见图 1-29），运行 PLC 观察运行状态。

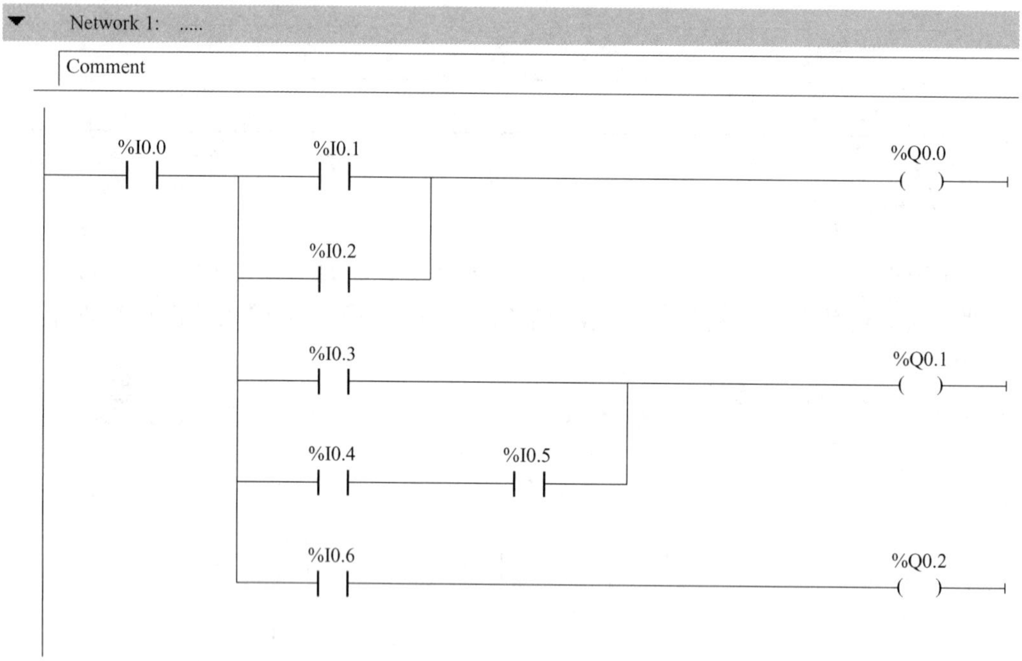

图 1-29　练习程序

项目 2　三相异步电动机的瞬时启停控制

三相异步电动机具有结构简单、制造方便、运行性能好、可节省各种材料、价格便宜等特点，使其在生产、生活中得到广泛应用。比如，车床、鼓风机、行车（吊车）、数控机床、提升机、电拖车、抽水机等，都离不开它。

三相异步电动机的控制可分为瞬时控制和延时控制两大类。本项目主要学习用两种不同方法对三相异步电动机进行点动自锁以及正反转控制：方法一是使用传统的继电器系统进行控制；方法二是使用德国西门子公司生产的 1200 系列 PLC 进行控制。本项目分为三项异步电动机的点动自锁控制和三相异步电动机的正反转控制两个任务，由浅入深逐步掌握继电器控制系统和 PLC 控制系统的使用。

任务 2.1　三相异步电动机的点动自锁控制

2.1.1　任务描述

本任务分为两个部分，分别为三相异步电动机的点动控制和自锁控制。每个部分都需要通过两种方法来实现相同的控制要求，即利用继电器控制系统和 PLC 控制系统分别完成对三相异步电动机的点动控制和自锁控制。

在继电器控制系统中完成电路图的绘制、电气元件的连接、通电前检测、通电后分析等步骤。在 PLC 控制系统中除完成上述要求外，还需要完成 PLC 的外部接线以及 PLC 程序的编写调试。最后总结任务实施的过程，并记录在相应表格中。

2.1.2　任务目标

（1）掌握三相异步电动机线路的接线、查线和操作。
（2）掌握三相异步电动机点动、自锁控制线路的原理。
（3）掌握短路保护、过载保护、欠压保护的原理。
（4）认识中间继电器触点的连接方法及所起到的作用。
（5）理解 PLC 控制点动和自锁的区别。
（6）掌握查找、分析和排除故障的能力。
（7）培养学生安全操作及团队意识。

2.1.3　任务相关知识

2.1.3.1　热继电器

电动机在运行过程中，如果负载过大，电动机的电流将超过它的额定值，若持续时间较长，电机的温升就会超过允许的温升值，将使电动机的绝缘损坏，甚至烧坏电动机。所以，对电动机过载需要采取保护措施。当电动机过载时，熔断器一般是不会熔断的，因为接于电动机主回路的熔断器主要用于电动机的短路保护，熔断器允许流过的电流值是电动机额定电流的好几倍。若熔断器容量选小了，电动机启动时就会经常使熔断器烧断。因此电动机过载保护需要采取别的措施，最常用的是采用热继电器进行过载保护。其实物图及电气符号分别如图 2-1 和图 2-2 所示。

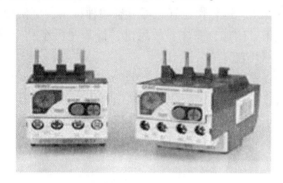

图 2-1　热继电器实物图

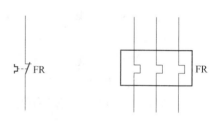

图 2-2　热继电器电气符号

热继电器的工作原理

热继电器是用于电动机或其他电气设备、电气线路的过载保护的保护电器。

电动机在实际运行中，如拖动生产机械进行工作过程中，若机械出现不正常的情况或电路异常使电动机遇到过载，则电动机转速下降、绕组中的电流将增大，使电动机的绕组温度升高。若过载电流不大且过载的时间较短，电动机绕组不超过允许温升，这种过载是允许的。但若过载时间长，过载电流大，电动机绕组的温升就会超过允许值，使电动机绕组老化，缩短电动机的使用寿命，严重时甚至会使电动机绕组烧毁。所以，这种过载是电动机不能承受的。热继电器就是利用电流的热效应原理，在出现电动机不能承受的过载时切断电动机电路，为电动机提供过载保护的保护电器。

使用热继电器对电动机进行过载保护时，将热元件与电动机的定子绕组串联，将热继电器的常闭触头串联在交流接触器的电磁线圈的控制电路中，并调节整定电流调节旋钮，使人字形拨杆与推杆相距适当距离。当电动机正常工作时，通过热元件的电流即为电动机的额定电流。热元件发热，双金属片受热后弯曲，使推杆刚好与人字形拨杆接触，而又不能推动人字形拨杆。常闭触头处于闭合状态，交流接触器保持吸合，电动机正常运行。

若电动机出现过载情况，绕组中电流增大，通过热继电器元件中的电流增大使双金属片温度升得更高，弯曲程度加大，推动人字形拨杆，人字形拨杆推动常闭触头，使触头断开而断开交流接触器线圈电路，使接触器释放、切断电动机的电源，电动机停车而得到保护。

热继电器的工作原理是电流入热元件的电流产生热量，使不同膨胀系数的双金属片发生形变。当形变达到一定距离时，就推动连杆动作，使控制电路断开，从而使接触器失电，主电路断开，实现电动机的过载保护。热继电器作为电动机的过载保护元件，以其体积小、结构简单、成本低等优点在生产中得到了广泛应用。

热继电器的日常维护

热继电器的日常维护包括：

(1) 热继电器动作后复位要一定的时间，自动复位时间应在 5min 内完成，手动复位要在 2min 后才能按下复位按钮。

(2) 当发生短路故障后，要检查热元件和双金属片是否变形。如有不正常情况，应及时调整，但不能将元件拆下。

(3) 使用中的热继电器每周应检查一次，具体内容包括：热继电器有无过热、异味及放电现象，各部件螺丝有无松动，脱落及解除不良，表面有无破损及清洁与否。

(4) 使用中的热继电器每年应检修一次，具体内容是：清扫卫生，查修零部件，测试绝缘电阻应大于 1MΩ，通电校验。经校验过的热继电器，除了接线螺钉之外，其他螺钉不要随便行动。

(5) 更换热继电器时，新安装的热继电器必须符合原来的规格与要求。

(6) 定期检查各接线电有无松动，在检修过程中绝不能折弯双金属片。

练一练

(1) 观察热继电器的外观，查看标签及技术参数；

(2) 思考如何接线，把端子对应的线号和功能写在表 2-1 中。

表 2-1　热继电器端子功能表

序号	端子名称	功能	序号	端子名称	功能
1			5		
2			6		
3			7		
4			8		

2.1.3.2　点动和自锁控制原理分析

通过 4 个部分对原理进行分析和介绍，其分别为继电器系统控制的点动线路、PLC 系统控制的点动线路、继电器系统控制的自锁线路和 PLC 系统控制的自锁线路。

继电器系统控制的点动线路

点动控制线路是用按钮、接触器来控制电动机的最简单的控制线路。点动控制是指按下按钮，电动机得电运转；松开按钮，电动机断电停转。这种控制方法常用于电动葫芦的起重电机和车床快速移动的电机控制。控制线路通常采用国家标准规定的电气图形符号和文字符号，画成控制线路原理图来表示。它是依据实物接线电路绘制的，用来表达控制线路的工作原理。其原理图如图 2-3 所示。

图 2-3 中的点动控制原理图可分成主电路和控制电路两大部分。主电路是从电源 L_1、L_2、L_3 经电源开关 QS、熔断器 FU_1、接触器 KM 的主触点到电动机 M 的电路，它流过的电路较大。由熔断器 FU_2、按钮 SB 和接触器 KM 的线圈组成控制电路，流过的电流较小。

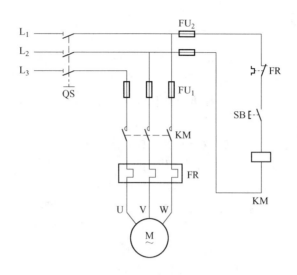

图 2-3 继电器系统控制的点动电路图

当电动机需点动时,先合上电源开关 QS,按下点动按钮 SB,接触器线圈 KM 便通电,衔铁吸合,带动它的三对常开主触点 KM 闭合,电动机 M 便接通电源起动运转。SB 按钮放开后,接触器线圈断电,衔铁受弹簧力的作用而复位,带动它的三对常开主触点断开,电动机断电停转。

点动控制线路的工作原理为:合上电源开关 QS 后,若要启动电动机,其操作步骤为:按下 SB→KM 线圈通电→KM 主触点闭合→电动机 M 运转;若要停止电动机,其操作步骤为:松开 SB→KM 线圈断电→KM 主触点断开→电动机 M 停转。

练一练

观察图 2-3:

(1) 自主分析其运行过程,并手绘电路图;

(2) 思考如何接线。

PLC 系统控制的点动线路

项目 1 中介绍了 PLC 的诸多优点,这里不再赘述。总之在引入了 PLC 进行控制后,可以增强系统的稳定性、提高运行速度等特点。其原理图如图 2-4 和图 2-5 所示。

由图 2-4 可以看出,采用中间继电器 KA 来当作弱电控制强电的桥梁,它可以对 PLC 起到保护的作用。即使有的 PLC 输出可以直接通交流电,但是为了安全起见,实际应用中往往采用中间继电器。本任务中采用的 KA 为直流 24V 供电,它包含一个线圈和一对常开触点。当按下 SB 时,PLC 得到输入信号,经内部运行计算后使 PLC 的输出有信号。KA 所在的线圈通电,使 KA 的常开触点闭合,导致 KM 的线圈通电。衔铁吸合,带动它的三对常开主触点 KM 闭合,电动机 M 便接通电源起动运转。FR 起到保护作用,可以考虑当作输入点接在 PLC 回路中。当松开 SB 时,PLC 输入信号中断,经内部运行计算后使 PLC 的输出有信号消失。KA 所在的线圈失电,使 KA 的常开触点恢复原来状态,导致 KM 的线圈失电。衔铁受弹簧力的作用而复位,带动它的三对常开主触点断开,电动机断电停转。

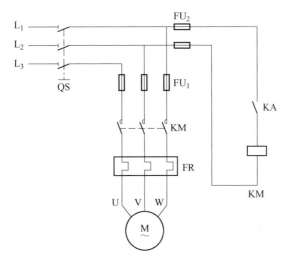

图 2-4 点动控制主电路接线图（PLC）

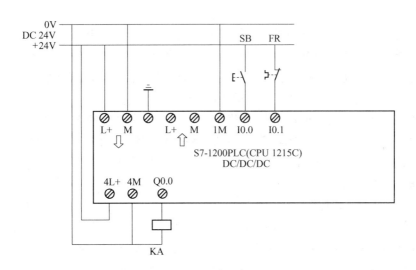

图 2-5 点动控制 PLC 控制电路接线图

练一练

观察图 2-4 和图 2-5：

（1）自主分析其运行过程，并手绘电路图；

（2）思考如何接线。

继电器系统控制的自锁线路

为了实现电动机的连续运行，可采用接触器自锁的正转控制线路，需要用接触器的一个常开辅助触点并联在起动按钮 SB_2 的两端，在控制电路中在串联一个停止按钮 SB_1，可以将电动机停止。这种接触器自锁的正转控制线路不但能使电动机连续运转，还具有欠压保护和失压（零压）保护的功能。其原理图如图 2-6 所示。

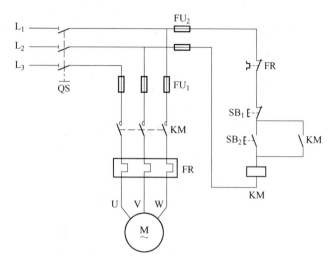

图 2-6 自锁控制继电器电路图

- 欠压保护

欠压是指线路电压低于电动机应加的额定电压。欠压保护是指线路电压低于某一数值时，电动机能自动脱离电源电压停转，避免电动机在欠压下运行的一种保护。电动机要有欠压保护是因为在电动机运行时，当电源电压下降，电动机的电流就会上升，电压下降越严重，电流上升的也越严重，甚至会烧坏电动机。

当电动机运转时，电源电压降低到较低（一般在工作电压的85%以下）时，接触器线圈的磁通变得很弱，电磁吸力不足，动铁心在反作用弹簧的作用下释放，自锁触点断开，失去自锁。同时主触点也断开，电动机停转，得到了保护。

- 失压（零压）保护

失压保护是指电动机运行时，由于外界某种原因使电源临时停电时，能自动切断电动机电源。在恢复供电时，而不能让电动机自行起动，如果未加防范措施很容易造成人身事故。由于自锁触点和主触点在停电时一起断开，采用接触器自锁的正转控制线路，控制电路和主电路都不会自行接通。所以在恢复供电时，如果没有按下按钮，电动机就不会自行启动。

- 过载保护

过载保护是指当电动机出现过载时能自动切断电动机电源，使电动机停转的一种保护。最常用的是利用热继电器进行过载保护。电动机在运行过程中，长期负载过大、操作频繁或断相运行等都可能使电动机定子绕组的电流超过它的额定值，但电流又未达到使熔断器熔断，将引起电动机定子绕组过热温度升高。如果温度超过允许温升，就会使绝缘损坏，电动机的使用寿命大幅度缩短，严重时甚至会烧坏电动机。因此，对电动机必须采取过载保护的措施。

电动机在运行过程中，由于过载或其他原因使电流超过额定值，经过一定时间，串接在主电路中的热继电器 FR 的热元件受热发生弯曲。因此需要通过动作机构使串接在控制电路中的 FR 常闭触点断开，切断控制电路，接触器 KM 的线圈断电，主触点断开，电动机 M 便停转，从而达到过载保护的目的。

练一练

观察自锁控制继电器电路图：

（1）自主分析其运行过程，并手绘电路图；

（2）思考如何接线。

PLC 系统控制的自锁线路

PLC 系统控制的自锁线路原理图如图 2-7 和图 2-8 所示。

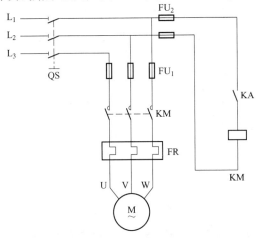

图 2-7　自锁控制主电路接线图（PLC）

由图 2-7 可以看出，可采用中间继电器 KA 当作弱电控制强电的桥梁。当按下 SB_1 时，PLC 得到输入信号，经内部运行计算后使 PLC 的输出有信号。KA 所在的线圈通电，使 KA 的常开触点闭合，导致 KM 的线圈通电。衔铁吸合，带动它的三对常开主触点 KM 闭合，电动机 M 便接通电源起动运转。FR 起到保护作用，可以考虑当作输入点接在 PLC 回路中。

当按下 SB_2 时，PLC 输入信号中断，经内部运行计算后使 PLC 的输出有信号消失。KA 所在的线圈失电，使 KA 的常开触点恢复原来状态，导致 KM 的线圈失电。衔铁受弹簧力的作用而复位，带动它的三对常开主触点断开，电动机断电停转。

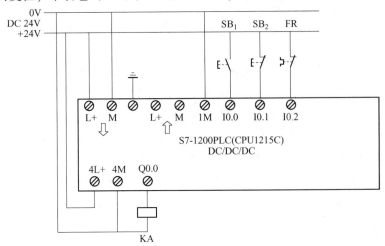

图 2-8　自锁控制 PLC 控制电路接线图

练一练

观察图 2-7 和图 2-8 电路图，

（1）自主分析其运行过程，并手绘电路图；

（2）思考如何接线。

2.1.3.3 输入输出指令的认知

输入指令多用于表示触点的状态，包括常开触点和常闭触点。输入指令多用于表示普通线圈的状态，包括得电状态和失电状态。它们是 PLC 编程中最基本的指令，也是学好 PLC 编程的入门条件。以下通过图 2-9 进行具体介绍。

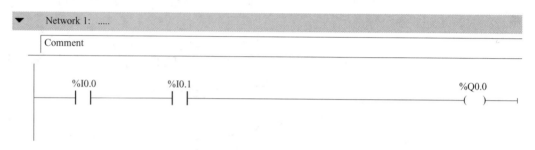

图 2-9 输入输出指令示例程序

输入指令

输入指令用于输入触点的状态，常开、常闭触点指令在梯形图（LAD）中是用触点表示的。如图 2-9 所示，I0.0 和 I0.1 都是输入指令，都表示常开触点，在梯形图（LAD）中，输入触点指令是指其触点与左母线连接。当触点为常闭状态时，能流可以通过触点；当触点为常开状态时，能流不能通过。

输出指令

输出指令是将逻辑运算结果写入输出映像寄存器中，最终决定输出端子状态。在梯形图（LAD）中，输出指令是以线圈的形式表示的。如图 2-9 所示，Q0.0 就是输出指令，它是一个普通线圈，表示得电或者失电的状态。

综合应用

Q0.0 表示的线圈同样也拥有 KM 线圈的特点，即线圈得电时，其对应的常开常闭触点动作；线圈失电时，其对应的常开常闭触点恢复原来的状态。如图 2-10 所示，前面的三个常开触点都闭合时，能流到达线圈处，使线圈 Q0.0 得电。由于线圈 Q0.0 的得电，导致 Q0.0 的常开触点动作，有常开状态变为常闭状态。这样一来，即使 I0.0 断开，Q0.0 也不会失电，除非 I0.1 或者 I0.2 断开，才能使 Q0.0 断电。

练一练

分析图 2-10，说明为什么 Q0.0 得电后，即使 I0.0 断开，Q0.0 也不会失电。

2.1.4 任务实施

在 XK-SX2C 高级维修电工实训台上，完成以下 4 个任务：继电器系统对三相异步电

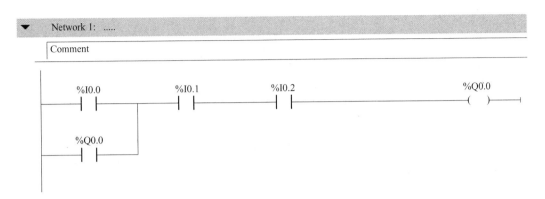

图 2-10　综合应用示例程序

动机点动的控制,PLC 系统对三相异步电机点动的控制,继电器系统对三相异步电动机自锁的控制,以及 PLC 系统对三相异步电动机自锁的控制。要求每个任务在确保接正确的基础上,接通电源。在实施过程中,如果出现故障,要对其进行故障的查找、分析和排除。注意安全意识和团队意识等职业核心素养的养成。

2.1.4.1　继电器系统对三相异步电动机点动的控制

任务要求

按照图 2-3 进行线路的连接,完成对三相异步电动机的点动控制。完成后对本项目进行评分,详见表 2-4。

操作步骤

其操作步骤分别为:

(1) 对所需元件进行故障查找,确保每个元件能正常使用。

(2) 绘制原理图,并标出线号。

(3) 按照绘制好的图纸进行线路连接。

(4) 用万用表做通电前的检查工作,若出现故障及时排除并记录总结。

(5) 通电观察系统的运行状态,若出现故障及时排除并记录总结。

(6) 把三相异步电动机接入线路中,通电观察其运行状态。

(7) 总结并记录。

分析总结

总结整个任务实施过程,查找不足,尤其对于出现的故障要高度重视,分析总结后记录在表 2-5 中。

2.1.4.2　PLC 系统对三相异步电动机点动的控制

任务要求

按照图 2-4 和图 2-5 进行线路的连接,完成对三相异步电动机的点动控制。完成后对本项目进行评分,详见表 2-4。

操作步骤

其操作步骤分别为：

（1）对所需元件进行故障查找，确保每个元件能正常使用。

（2）绘制原理图，并标出线号。

（3）连接 PLC 的电源线，用万用表测量，确保无误后方可通电。

（4）确定 I/O 地址分配，并填入表 2-2 中。

表 2-2　I/O 地址分配表

输入信号				输出信号			
序号	功能	元件	地址	序号	控制对象	元件	地址
1				1			
2				2			
3				3			

（5）根据 I/O 地址分配完成输入端的线路连接。

（6）检查无误后再给 PLC 通电。

（7）编写程序，下载到 PLC 中进行调试。

（8）按照绘制好的图纸完成整个线路的连接。

（9）用万用表做通电前的检查工作，若出现故障及时排除并记录总结。

（10）通电观察系统的运行状态，若出现故障及时排除并记录总结。

（11）把三相异步电动机接入线路中，通电观察其运行状态。

（12）总结并记录。

分析总结

总结整个任务实施过程，查找不足，尤其对于出现的故障要高度重视，分析总结后记录在表 2-5 中。

2.1.4.3　继电器系统对三相异步电动机自锁的控制

任务要求

按照图 2-6 进行线路的连接，完成对三相异步电动机的自锁控制。完成后对本项目进行评分，详见表 2-4。

操作步骤

其操作步骤分别为：

（1）对所需元件进行故障查找，确保每个元件能正常使用。

（2）绘制原理图，并标出线号。

（3）按照绘制好的图纸进行线路连接。

（4）用万用表做通电前的检查工作，若出现故障及时排除并记录总结。

（5）通电观察系统的运行状态，若出现故障及时排除并记录总结。

（6）把三相异步电动机接入线路中，通电观察其运行状态。

（7）总结并记录。

分析总结

总结整个任务实施过程，查找不足，尤其对于出现的故障要高度重视，分析总结后记录在表2-5中。

2.1.4.4 继电器系统对三相异步电动机自锁的控制

任务要求

按照图2-7和图2-8进行线路的连接，完成对三相异步电动机的自锁控制。完成后对本项目进行评分，详见表2-4。

操作步骤

其操作步骤分别为：

（1）对所需元件进行故障查找，确保每个元件能正常使用。
（2）绘制原理图，并标出线号。
（3）连接PLC的电源线，用万用表测量，确保无误后方可通电。
（4）确定I/O地址分配，并填入表2-3中。

表2-3 I/O地址分配表

输入信号				输出信号			
序号	功能	元件	地址	序号	控制对象	元件	地址
1				1			
2				2			
3				3			

（5）根据I/O地址分配完成输入端的线路连接。
（6）检查无误后再给PLC通电。
（7）编写程序，下载到PLC中进行调试。
（8）按照绘制好的图纸完成整个线路的连接。
（9）用万用表做通电前的检查工作，若出现故障及时排除并记录总结。
（10）通电观察系统的运行状态，若出现故障及时排除并记录总结。
（11）把三相异步电动机接入线路中，通电观察其运行状态。
（12）总结并记录。

分析总结

总结整个任务实施过程，查找不足，尤其对于出现的故障要高度重视，分析总结后记录在表2-5中。

2.1.5 任务评价

对第4个任务——继电器系统对三相异步电动机自锁的控制进行评价，并将评分结果分别填入表2-4内。

表 2-4 评分表

任务内容	考核要求	评分标准	配分	扣分	得分
准备工作	(1) 对所需元件进行故障查找，确保每个元件能正常使用； (2) 绘制原理图，并标出线号	准备工作是整个实施过程的入门阶段，这个阶段完成后才有资格进行下一步，如果这步未做或者失败，整个项目记 0 分	10		
工作原理描述	能够正确的说明整个系统的工作原理	(1) 主电路功能描述 5 分； (2) PLC 线路功能描述 5 分	10		
PLC 控制系统主控电路接线	正确完成 PLC 主电路及控制电路的接线	(1) 每接错一根线扣 5 分； (2) 其他情况酌情扣分	20		
PLC 程序编写与下载	按照任务要求将 PLC 程序输入、下载与监控	(1) 软件使用不熟练，不能完成程序的输入操作扣 15 分； (2) 不会进行程序的 IP 地址设置与下载扣 5 分； (3) 不能通过博途软件控制 PLC 的停止和运行状态扣 5 分； (4) 不会操作博途软件监控程序的运行扣 5 分	20		
PLC 控制系统运行演示	正确演示电动机的启动和停止，并能结合程序、硬件进行说明	(1) 电路不能实现启动扣 10 分； (2) 电路不能实现停止扣 10 分； (3) 不能对操作现象进行分析和说明的每处扣 5 分	30		
安全文明生产	遵守 8S 管理制度，遵守安全管理制度	(1) 未穿戴劳动保护用品扣 10 分； (2) 操作存在安全隐患，每次扣 5 分，扣完为止； (3) 操作现场未及时整理整顿，每次发现扣 2 分，扣完为止	10		
总 分					

2.1.6 任务小结

任务分析总结记录单见表 2-5。

表 2-5　任务分析总结记录单

任务分析总结记录单			
故障 1	故障现象		
	故障原因		
	排除过程		
故障 2	故障现象		
	故障原因		
	排除过程		
故障 3	故障现象		
	故障原因		
	排除过程		
总　结			

2.1.7　任务拓展

任务要求：分别使用两种控制系统实现两台三相异步电动机的顺序启停控制。

要求：两台电动机都有自己的启动和停止按钮。第一台电动机启动后，第二台电动机才能启动；若两台电动机都处在运行状态，第二台电动机停止后，第一台电动机才能停止。

任务 2.2　三相异步电动机的正反转控制

2.2.1　任务描述

本任务为三相异步电动机的正反转控制，要求用两种方法来实现相同的控制要求，即利用继电器控制系统和 PLC 控制系统分别完成对三相异步电动机的正反转控制。

在继电器控制系统中完成电路图的绘制、电气元件的连接、通电前检测、通电后分析等步骤。在 PLC 控制系统中除完成上述要求外，还需要完成 PLC 的外部接线以及 PLC 程序的编写调试。最后总结任务实施的过程，并记录在相应表格中。

2.2.2　任务目标

（1）理解互锁的含义、作用以及实现互锁的方法。

（2）学会实现电机正反转的各种方法以及注意事项。

（3）掌握三相异步电动机正反转线路的原理。

（4）理解接触器互锁正反转和双重联锁正反转的区别。

（5）掌握查找、分析和排除故障的能力。

（6）培养学生安全操作及团队意识。

2.2.3 任务相关知识

2.2.3.1 电动机正反转基础知识

电机正反转代表的是电机顺时针转动和逆时针转动。电机顺时针转动是电机正转，电机逆时针转动是电机反转。正反转控制电路图及其原理分析要实现电动机的正反转，只要将接至电动机三相电源进线中的任意两相对调接线即可达到反转的目的。电机的正反转在机械加工中广泛使用，例如行车、木工用的电刨床、台钻、刻丝机、甩干机和车床等。

最初人们需要某种设备反转需要将电机导线拆换，但这种方法在实际使用中烦琐。随着接触器的诞生，电机的正反转电路也有了进一步的发展，可以更加灵活方便地控制电机的正反转，并且在电路中增加了保护电路—互锁和双重互锁。可以实现低电压和远距离频繁控制。

为了使电动机能够正转和反转，可采用两只接触器 KM_1、KM_2 换接电动机三相电源的相序。但两个接触器不能同时吸合，如果同时吸合将造成电源的短路事故。为了防止这种事故，在电路中应采取可靠的互锁，因此需要采用按钮互锁、接触器互锁、按钮和接触器双重互锁的电动机正、反两方向运行的控制电路。

2.2.3.2 正反转线路控制原理分析

接触器互锁正反转控制线路

如图 2-11 所示，线路中采用了两个接触器，即正转接触器 KM_1 和反转接触器 KM_2。它们分别由正转按钮 SB_2 和反转按钮 SB_3 控制。当 KM_1 主触点接通时，三相电源 L_1、L_2、L_3、按 U—V—W 相序接入电动机；当 KM_2 主触点接通时，三相电源 L_1、L_2、L_3、按 W—V—U 相序接入电动机、即 W 和 U 两相相序反了一下。相应地控制电路有两条：一条是由正转按钮 SB_2 和 KM_1 线圈等组成的正转控制电路；另一条是由反转按钮 SB_3 和 KM_2 线圈等组成的反转控制电路。所以当两只接触器分别工作时，电动机的旋转方向相反。必须指出的是，线路要求接触器 KM_1 和 KM_2 不能同时接电，否则它们的主触点同时闭合，将造成 L_1、L_2、L_3、两相电源短路。为此，在接触器 KM_1 和 KM_2 线圈各自的支路中相互串联了对方的一对常闭辅助触点，即在正转控制电路串接反转接触器 KM_2 的常闭辅助触点，在反转控制电路串接了正转接触器 KM_1 的常闭辅助触点，从而保证接触器 KM_1 和 KM_2 不会同时通电。KM_1 与 KM_2 的这两对常闭辅助触点在线路中所起的作用称作互锁（或联锁），这两对触点称作互锁触点（或联锁触点）。

如图 2-11 所示是电动机正反转控制的一种典型线路，但这种线路要改变电动机的转向时，必须先按停止按钮 SB_1，再按反转按钮 SB_3，操作不方便。

按钮互锁正反转控制线路

为克服接触器联锁正反转控制线路操作不便，把正转按钮 SB_2 和反转按钮 SB_3 换成两个复合按钮，并使两个复合按钮的常闭触头代替接触器的联锁触头，从而构成了按钮联锁的正反转控制线路，如图 2-12 所示。

这种控制线路的工作原理与接触器联锁的正反转控制线路的工作原理基本相同，只是

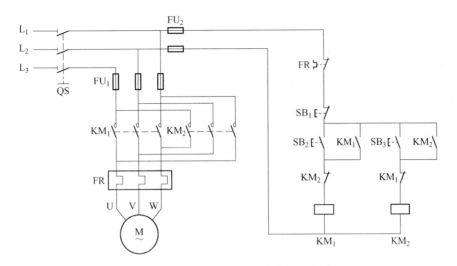

图 2-11 接触器互锁正反转控制线路图

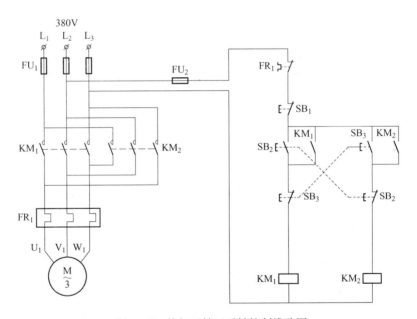

图 2-12 按钮互锁正反转控制线路图

当电动机从正转变为反转时,可直接按下反转按钮 SB_3 即可实现,不必先按停止按钮 SB_1。这样既保证了 KM_1 和 KM_2 的线圈不会同时通电,又可不按停止按钮而直接按反转按钮实现反转。同样,若使电动机从反转运行变为正转运行时,只需直接按下正转按钮 SB_2 即可。

这种线路的优点是操作方便,缺点是容易产生电源两相短路故障。例如,当正转接触器 KM_1 发生主触头熔焊或被杂物卡住等故障时,即使 KM_1 线圈失电,主触头也分断不开,这时若直接按下反转按钮 SB_2,KM_2 得电动作,触头闭合,必然造成电源两相短路故障。所以采用此线路工作有一定安全隐患。在实际工作中,经常采用按钮、接触器双重联锁的正反转控制线路。

按钮、接触器双重联锁的正反转控制线路

为克服接触器联锁正反转控制线路和按钮联锁正反转控制线路的不足，在按钮联锁的基础上，又增加了接触器联锁，构成按钮、接触器双重联锁正反转控制线路，如图 2-13 所示。该线路兼有两种联锁控制线路的优点，操作方便工作安全可靠。

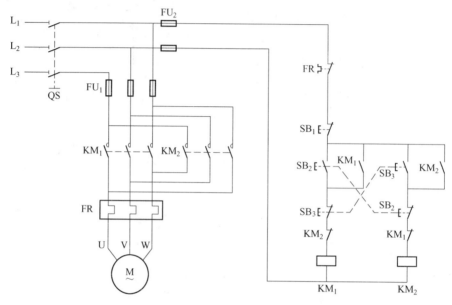

图 2-13　按钮、接触器双重联锁的正反转控制线路图

练一练

（1）分析讨论以上三种正反转的控制线路的原理，说明为什么 KM_1 和 KM_2 不能同时吸合。

（2）画出以上三种控制线路图，标出线号，并思考如何接线。

2.2.3.3　PLC 系统控制的正反转线路分析

由于按钮互锁正反转控制线路有缺点，基本不被采用，所以只介绍剩余两种正反转控制方式。由于 PLC 的加入，大大简化了控制回路，使得接触器互锁控制线路和双重互锁控制线路的硬件接线图基本一样（见图 2-14 和图 2-15），唯一的区别只是程序的不同。

线路图

采用中间继电器 KA 当作弱电控制强电的桥梁，继电器 KA 可以对 PLC 起到保护的作用。即使有的 PLC 输出可以直接通交流电，但是为了安全起见，实际应用中往往采用中间继电器。当按下 SB_1 时，PLC 得到输入信号，经内部运行计算后使 PLC 的输出有信号。KA_1 所在的线圈通电，使 KA_1 的常开触点闭合。导致 KM_1 的线圈通电。衔铁吸合，带动它的三对常开主触点 KM_1 闭合，电动机正转运转。SB_2 的作用与 SB_1 类似，可自行分析。FR 起到保护作用，可以考虑当作输入点接在 PLC 回路中。SB_3 为系统的停止按钮。

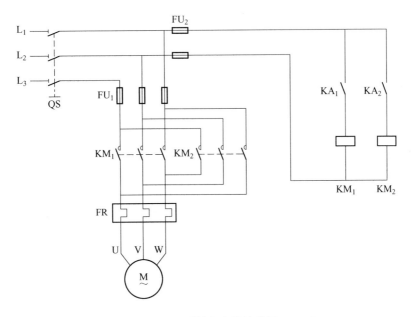

图 2-14 正反转主电路接线图（PLC）

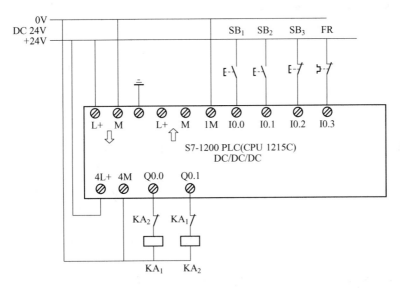

图 2-15 正反转 PLC 控制电路接线图

互锁程序的编写

参考程序如图 2-16 和图 2-17 所示，其分别为接触器互锁正反转和双重连锁正反转的程序。

练一练

（1）分析图 2-14 和图 2-15：

1）自主分析其运行过程，并手绘电路图；

2）思考如何接线。

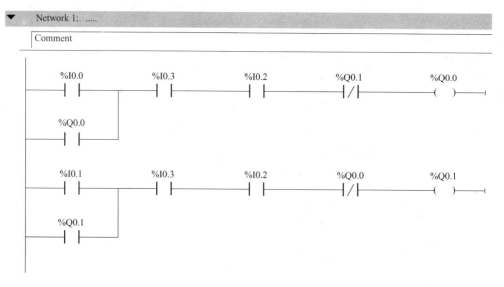

图 2-16 接触器互锁正反转参考程序

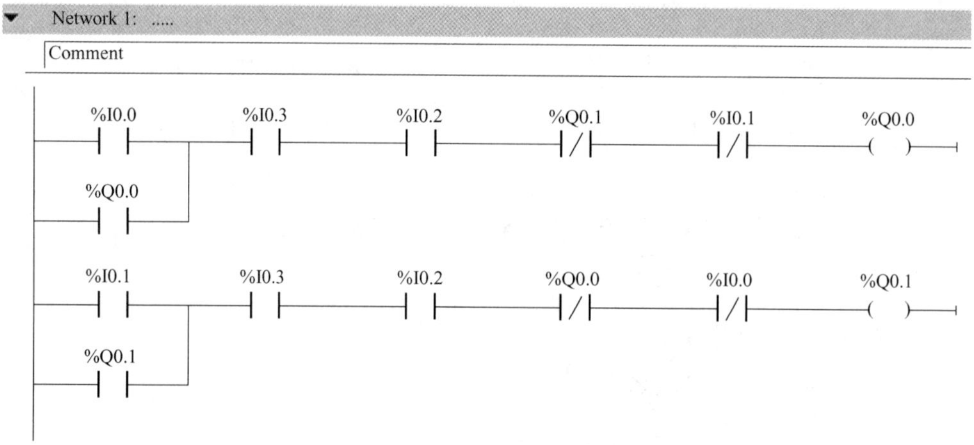

图 2-17 双重联锁正反转参考程序

（2）把示例程序下载到 PLC 中，观察 PLC 的运行状态。

（3）思考如何编写 PLC 控制按钮、接触器双重联锁的正反转线路程序。

2.2.4 任务实施

在 XK-SX2C 高级维修电工实训台上，完成以下 4 个任务：继电器系统对三相异步电动机接触器互锁正反转的控制，PLC 系统对三相异步电动机接触器互锁正反转的控制，继电器系统对三相异步电动机双重联锁正反转的控制，以及 PLC 系统对三相异步电动机双重联锁正反转的控制。要求每个任务在确保连接正确的基础上，接通电源。在实施过程中，如果出现故障，要对其进行故障的查找、分析和排除。注意安全意识和团队意识等职业核心素养的养成。

2.2.4.1 继电器系统对三相异步电动机接触器互锁正反转的控制

任务要求

按照图 2-14 和图 2-15 进行线路的连接,完成对三相异步电动机的接触器互锁正反转控制。完成后对本项目进行评分,详见表 2-8。

操作步骤

其操作步骤分别为:

(1) 对所需元件进行故障查找,确保每个元件能正常使用。
(2) 绘制原理图,并标出线号。
(3) 按照绘制好的图纸进行线路连接。
(4) 用万用表做通电前的检查工作,若出现故障及时排除并记录总结。
(5) 通电观察系统的运行状态,若出现故障及时排除并记录总结。
(6) 把三相异步电动机接入线路中,通电观察其运行状态。
(7) 总结并记录。

分析总结

总结整个任务实施过程,查找不足,尤其对于出现的故障要高度重视,分析总结后记录在表 2-9 中。

2.2.4.2 PLC 系统对三相异步电动机接触器互锁正反转的控制

任务要求

按照图 2-14 和图 2-15 进行线路的连接,完成对三相异步电动机的接触器互锁正反转控制。完成后对本项目进行评分,详见表 2-8。

操作步骤

其操作步骤分别为:

(1) 对所需元件进行故障查找,确保每个元件能正常使用。
(2) 绘制原理图,并标出线号。
(3) 连接 PLC 的电源线,用万用表测量,确保无误后方可通电。
(4) 确定 I/O 地址分配,并填入表 2-6 中。

表 2-6 I/O 地址分配表

输入信号				输出信号			
序号	功能	元件	地址	序号	控制对象	元件	地址
1				1			
2				2			
3				3			
4				4			

(5) 根据 I/O 地址分配完成输入端的线路连接。

(6) 检查无误后再给 PLC 通电。

(7) 编写程序，下载到 PLC 中进行调试。

(8) 按照绘制好的图纸完成整个线路的连接。

(9) 用万用表做通电前的检查工作，若出现故障及时排除并记录总结。

(10) 通电观察系统的运行状态，若出现故障及时排除并记录总结。

(11) 把三相异步电动机接入线路中，通电观察其运行状态。

(12) 总结并记录。

分析总结

总结整个任务实施过程，查找不足，尤其对于出现的故障要高度重视，分析总结后记录在表 2-9 中。

2.2.4.3 继电器系统对三相异步电动机双重联锁正反转的控制

任务要求

按照图 2-14 和图 2-15 进行线路的连接，完成对三相异步电动机的双重联锁正反转控制。完成后对本项目进行评分，详见表 2-8。

操作步骤

其操作步骤分别为：

(1) 对所需元件进行故障查找，确保每个元件能正常使用。

(2) 绘制原理图，并标出线号。

(3) 按照绘制好的图纸进行线路连接。

(4) 用万用表做通电前的检查工作，若出现故障及时排除并记录总结。

(5) 通电观察系统的运行状态，若出现故障及时排除并记录总结。

(6) 把三相异步电动机接入线路中，通电观察其运行状态。

(7) 总结并记录。

分析总结

总结整个任务实施过程，查找不足，尤其对于出现的故障要高度重视，分析总结后记录在表 2-9 中。

2.2.4.4 PLC 系统对三相异步电动机双重联锁正反转的控制

任务要求

按照图 2-14 和图 2-15 进行线路的连接，完成对三相异步电动机的双重联锁正反转控制。完成后对本项目进行评分，详见表 2-8。

操作步骤

其操作步骤分别为：

(1) 对所需元件进行故障查找，确保每个元件能正常使用。

(2) 绘制原理图，并标出线号。

(3) 连接 PLC 的电源线，用万用表测量，确保无误后方可通电。

(4) 确定 I/O 地址分配，并填入表 2-7 中。

表 2-7 I/O 地址分配表

输入信号				输出信号			
序号	功能	元件	地址	序号	控制对象	元件	地址
1				1			
2				2			
3				3			
4				4			

(5) 根据 I/O 地址分配完成输入端的线路连接。
(6) 检查无误后再给 PLC 通电。
(7) 编写程序，下载到 PLC 中进行调试。
(8) 按照绘制好的图纸完成整个线路的连接。
(9) 用万用表做通电前的检查工作，若出现故障及时排除并记录总结。
(10) 通电观察系统的运行状态，若出现故障及时排除并记录总结。
(11) 把三相异步电动机接入线路中，通电观察其运行状态。
(12) 总结并记录。

分析总结

总结整个任务实施过程，查找不足，尤其对于出现的故障要高度重视，分析总结后记录在表 2-9 中。

2.2.5 任务评价

对第四个任务——继电器系统对三相异步电动机自锁的控制进行评价，并将评分结果分别填入表 2-8 内。

表 2-8 评分表

任务内容	考核要求	评分标准	配分	扣分	得分
准备工作	(1) 对所需元件进行故障查找，确保每个元件能正常使用；(2) 绘制原理图，并标出线号	准备工作是整个实施过程的入门阶段，这个阶段完成后才有资格进行下一步，如果这步未做或者失败，整个项目记 0 分	10		
工作原理描述	能够正确的说明整个系统的工作原理	(1) 主电路功能描述 5 分；(2) PLC 线路功能描述 5 分	10		
PLC 控制系统主控电路接线	正确完成 PLC 主电路及控制电路的接线	(1) 每接错一根线扣 5 分；(2) 其他情况酌情扣分	20		

续表 2-8

任务内容	考核要求	评分标准	配分	扣分	得分
PLC 程序编写与下载	按照任务要求将 PLC 程序输入、下载与监控	(1) 软件使用不熟练，不能完成程序的输入操作扣 15 分； (2) 不会进行程序的 IP 地址设置与下载扣 5 分； (3) 不能通过博途软件控制 PLC 的停止和运行状态扣 5 分； (4) 不会操作博途软件监控程序的运行扣 5 分	20		
PLC 控制系统运行演示	正确演示电动机的启动、停止和正反转，并能结合程序、硬件进行说明	(1) 电路不能实现启动扣 10 分； (2) 电路不能实现停止扣 10 分； (3) 你能实现正反转功能扣 10 分； (4) 不能对操作现象进行分析和说明的每处扣 5 分	30		
安全文明生产	遵守 8S 管理制度，遵守安全管理制度	(1) 未穿戴劳动保护用品扣 10 分； (2) 操作存在安全隐患，每次扣 5 分，扣完为止； (3) 操作现场未及时整理整顿，每次发现扣 2 分，扣完为止	10		
总　　分					

2.2.6　任务小结

任务分析总结记录单见表 2-9。

表 2-9　任务分析总结记录单

任务分析总结记录单		
故障 1	故障现象	
	故障原因	
	排除过程	
故障 2	故障现象	
	故障原因	
	排除过程	
故障 3	故障现象	
	故障原因	
	排除过程	
总　结		

2.2.7 任务拓展

有些生产机械，如万能铣床，要求工作台在一定距离内能自动往复，不断循环，以便工件能连续加工，提高生产效率。工作台自动往复移动的示意图如图 2-18 所示。工作台上装有挡铁 1 和 2，机床床身上装有行程开关 SQ_1 和 SQ_2。当挡铁碰撞行程开关后，自动换接电动机正反转控制电路，使工作台自动往返移动。工作台的行程可通过移动挡铁的位置来调节，以适应加工零件的不同要求。SQ_3 和 SQ_4 用来做限位保护，即限制工作台的极限位置，以防止 SQ_1、SQ_2 失灵，工作台越过限定位置而造成事故。

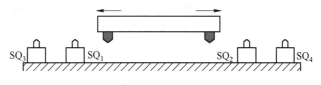

图 2-18 工作台自动往复移动示意图

位置开关（又名行程开关、限位开关）是一种将机械信号转换为电气信号以控制运动部件位置或行程的控制电器。位置控制线路就是用运动部件上的挡铁碰撞位置开关，而使其触点动作，以接通或断开电路来控制机械行程，或实现加工过程的自动往返。线路简单不受各种参数影响，只反映运动部件的位置。

2.2.8 项目小结

本项目包含两个任务，三相异步电动机的点动自锁控制和正反转控制都是瞬时控制，没有延时功能。本项目是实际工业控制的入门技术，该知识技能一定要掌握，特别是对互锁知识的理解和运用能力。互锁共分为三种方式，分别为按钮互锁、接触器互锁和双重联锁。按钮互锁可靠程度低，但是可以实现不停车改变运行方向；接触器互锁可靠性高，但是需要停车才能改变运行方向；双重联锁结合了前两种的优点，可靠性高，改变运行方向时不需要停车。

练习题

(1) 画出按钮和接触器双重联锁正反转控制电路图。
(2) 三相异步电动机互锁方式有几种？分析其优缺点。
(3) 编写一个三组抢答器的程序，每组都有一个抢答按钮和一个指示灯，谁先按下抢答按钮谁的指示灯点亮，后按下去的指示灯就没有反应。
(4) 在第(3)题的基础上增加一名裁判，裁判手中有开始和复位按钮，当裁判按下开始按钮后才可以抢答，裁判按下复位按钮后本次答题结束。

项目3　三相异步电动机的延时控制

电动机的延时控制在生产和生活中应用十分普遍，如楼宇电梯的升降及电梯门的开关动作、皮带机输送系统的启动和停止等。继电器控制电路实现电动机延时控制是通过时间继电器来实现的，而对于 PLC 控制电路，则是通过 PLC 内部的定时器指令来实现对电动机的各种延时控制。本项目主要学习如何使用时间继电器和定时器指令来设计相应的主电路和控制电路，以及控制电动机的延时运行。由浅入深讲解电动机延时控制的主电路和控制电路的设计。

任务3.1　电动机的延时启动主电路及控制电路的设计与实现

3.1.1　任务描述

本任务分别利用继电器系统和 PLC 控制系统完成对三相异步电动机的延时启动主控电路的设计、接线和调试。

3.1.2　任务目标

（1）掌握时间继电器的使用方法。
（2）掌握三相异步电动机的延时启动继电器控制系统的主控电路的设计、接线和调试。
（3）理解西门子 S7-1200 PLC 定时器指令的工作原理。
（4）掌握三相异步电动机的延时启动 PLC 控制系统的主控电路的设计和接线。
（5）能够独立完成 PLC 程序的设计与调试。

3.1.3　任务相关知识

3.1.3.1　时间继电器的认知

时间继电器是一种利用电磁原理或机械动作原理实现触头延时接通或断开的自动控器。根据触头延时的特点，时间继电器可分为通电延时和断电延时两种。根据工作原理不同，时间继电器又可分为空气阻尼式、电子式、数显式和电磁式。其中，电子式时间继电器应用较为普遍。本项目使用的时间继电器为电子式，其实物图如图 3-1 所示。

图 3-1　电子式时间继电器实物图

电子式时间继电器的线圈及触点的电气符号如图 3-2 所示。

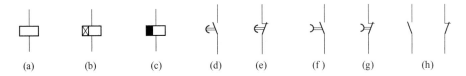

图 3-2　电子式时间继电器电气符号

在图 3-2 中，(a)为时间继电器线圈的一般形式；(b)为通电延时继电器线圈；(c)为断电延时继电器线圈；(d)和(e)为通电延时继电器的延时触点；(f)和(g)为断电延时继电器的延时触点；(h)为时间继电器的瞬动触点。不论是线圈还是触点，时间继电器字母符号均为 KT。

将时间继电器从底座上拔出，可以看到每个端子对应一个端子号，对照时间继电器旁边印刷的接线图如图 3-3 所示。其中，②和⑦为电压输入端，①、④、⑤和⑧为常闭触头，①、③、⑧和⑥为常开触头。接线完毕后，将时间继电器插入底座。

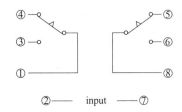

图 3-3　电子式时间继电器接线示意图

练一练

实际调整一下时间继电器的时间设置，使用万用表找出时间继电器的闭合和断开触点。

3.1.3.2　定时器指令

S7-1200 系列 PLC 是采用 IEC 标准的定时器指令，用户程序中可以使用的定时器数仅受 CPU 存储器容量限制，每个定时器均使用 16 个字节的 IEC_TIMER 数据类型的 DB 结构来存储功能框或线圈指令顶部指定的定时器数据。S7-1200 系列 PLC 的定时器种类有脉冲型定时器、接通延时定时器、断开延时定时器和保持性接通延时定时器。这里仅对脉冲型定时器和接通延时定时器进行介绍。

脉冲型定时器

脉冲型定时器的指令梯形图形式如图 3-4(a)所示，其标识符为 TP。该指令用于可生成具有预设宽度时间的脉冲，定时器指令的 IN 管脚用于启用定时器，PT 管脚表示定时器的设定值，Q 表示定时器的输出状态，ET 表示定时器的当前值。脉冲型定时器的指令格式和定时器指令执行时的时序图如图 3-4(b)所示。

注：数据块 DB1 为系统自动分配。

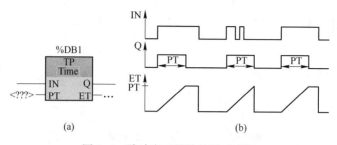

图 3-4 脉冲定时器符号及时序图

使用 TP 指令，可以将输出 Q 置位预设一段时间，当定时器使能端的状态从 OFF 变为 ON 时，可启动该定时器指令，定时器开始计时。无论后续使能端的状态如何变化，都可将输出 Q 置位由 PT 指定一段时间。若定时器正在计时，即使检测到使能端的信号在此从 OFF 变为 ON 的状态，输出 Q 的信号状态也不会受到影响。

根据脉冲型定时器的时序图分析出如下程序执行过程如图 3-5 所示。

图 3-5 脉冲定时器举例

当 I0.5 接通为 ON 时，Q0.4 的状态为 ON，5s 后，Q0.4 的状态变为 OFF。在这 5s 内，不管 I0.5 的状态如何变化，Q0.4 的状态始终保持为 ON。

接通延时定时器

接通延时定时器的指令标识符为 TON，如图 3-6(a) 所示。接通延时定时器输出端 Q 在预设的延时时间过后，输出状态为 ON，指令中管脚的定义与 TP 定时器指令管脚的定义一致。图 3-6(b) 描述了接通延时定时器的指令格式及执行时序图。

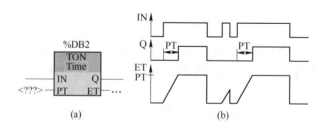

图 3-6 通延时器符号及时序图

当定时器的使能端 IN 为 1 时，该指令启动。定时器指令启动后开始计时，在定时器的当前值 ET 与设定值 PT 相等于时，输出端 Q 输出为 ON。只要使能端的状态仍为 ON，输出端 Q 就保持输出为 ON。若使能端的信号状态变为 OFF，则将复位输出端 Q 为 OFF。在使能端再次变为 ON 时，该定时器功能将再次启动。接通延时定时器的工作实例如图 3-7 所示。

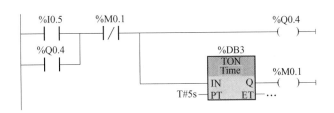

图 3-7 接通延时定时器举例

该段程序主要完成的是启动输出后,延时一段时间后自动断开的程序。当 I0.5 接通为 ON 时,Q0.4 得电自锁输出为 ON。当 Q0.4 输出为 ON 时,启动接通延时定时器 TON,使该定时器工作进行延时。延时 5s 后,定时器的输出端 Q 输出为 ON 状态,使 M0.1 线圈得电为 ON,从而使程序中 M0.1 的常闭触点断开,Q0.4 自锁解除断开为 OFF。

练一练

从博图软件的指令树中分别找出通电延时定时器指令和脉冲定时器指令,将它们分别拖拽到编程界面中加以熟悉。

3.1.4 任务实施

3.1.4.1 电动机的延时启动主控电路的接线与调试(继电器控制系统)

任务要求

点动启动按钮,电动机延时 10s 启动;点动停止按钮,电动机立即停止。完成主电路和控制电路的接线及调试,电路应具备必要的短路和过载保护措施。

完成后对本项目进行评分,详见表 3-2。

任务分析

按下启动按钮,时间继电器通电并开始计时。若断开启动按钮,且时间继电器仍然通电,则需要借助中间继电器的自锁实现。定时时间到后,时间继电器的延时闭合触点闭合,使控制电动机工作的接触器线圈自锁,电动机启动运行。此时,时间继电器和中间继电器任务已经完成,通过接触器的常闭触点将它们所在电路断开。主电路应有熔断器和热继电器;控制电路应有熔断器;以及热继电器的常闭触点。

综上所述,使用时间继电器实现电动机延时启动主控电路如图 3-8 所示。

操作步骤

其操作步骤分别为:

(1)在 XK-SX2C 高级维修电工实训台上,使用跨接线完成如图 3-8 所示的主电路和控制电路连接。

(2)设置时间继电器的动作时间为 10s。

(3)使用万用表分别测量主电路和控制电路,检查电路是否有短路、断路或接错线。如果存在问题,仔细检查,直到排除故障。

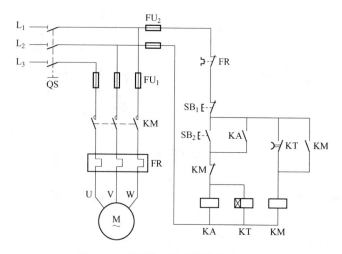

图 3-8 延时启动继电器主控电路图

任务演示

其演示步骤分别为：

（1）闭合实训台总电源和电路隔离开关 QS。

（2）点动 SB_2，中间继电器线圈吸合，时间继电器指示灯亮，10s 后接触器线圈吸合，电动机启动，同时中间继电器和时间继电器断电。

（3）点动 SB_1，接触器线圈断电释放，主触点断开。电动机立即停止运行。

（4）断开隔离开关 QS 和实训台总电源。

3.1.4.2 电动机的延时启动主控电路的接线与调试（PLC 控制系统）

任务要求

点动与 PLC 输入点连接的启动按钮，电动机延时 10s 启动；点动与 PLC 输入点连接停止按钮，电动机立即停止。完成主电路和 PLC 控制电路的接线及调试，电路应具备必要的短路和过载保护措施。

完成后对本项目进行评分，详见表 3-2。

任务分析

设计 PLC 控制系统，首先应根据设计要求进行分析，确定 PLC 的 I/O 地址分配，电动机延时启动 PLC 控制系统 I/O 地址分配表见表 3-1。

表 3-1 电动机延时启动 PLC 控制系统 I/O 地址分配表

输入信号				输出信号			
序号	功能	元件	地址	序号	控制对象	元件	地址
1	启动按钮	SB1	I0.0	1	中间继电器	KA	Q0.0
2	停止按钮	SB2	I0.1	—	—	—	—
3	过载保护	FR	I0.2	—	—	—	—

在 PLC 的地址分配表基础上，设计 PLC 的外部接线图如图 3-9 所示。

对接触器的控制仍然需要通过中间继电器的常开触点来控制接触器线圈（即弱电控制强电的方式）实现，如图 3-10 所示。

在主电路和控制电路设计完成的基础上，要进行 PLC 程序的设计，参考程序如图 3-11 所示。

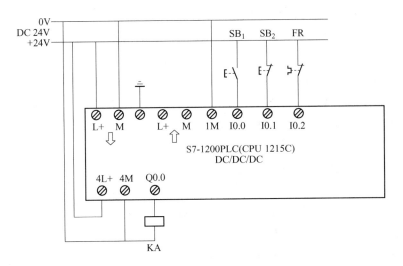

图 3-9 延时启动 PLC 控制电路接线图

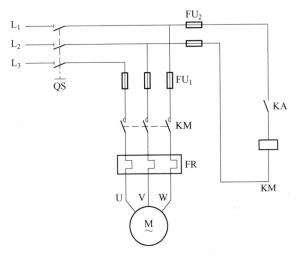

图 3-10 电动机延时启动主电路接线图（PLC）

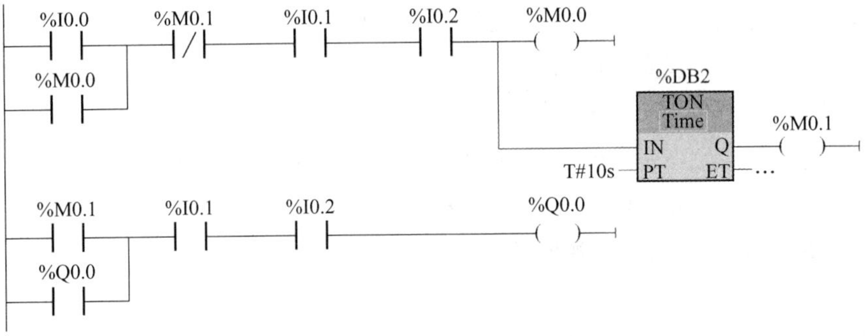

图 3-11 电动机延时启动 PLC 参考程序

· 211 ·

程序执行及硬件设备运行情况如下:

(1) 延时启动:点动 SB_1,I0.0 接通后断开,使 M0.0 线圈自锁,定时器 IN 端接通使其开始计时。当定时器计时到 10s 时,定时器 Q 端接通,M0.1 线圈接通。由于 M0.1 线圈接通,其对应的两个触点动作会发生:

1) M0.1 常闭触点断开,使得 M0.0 线圈自锁破坏,M0.0 线圈失电。定时器 IN 端由于断开而停止计时,定时器 Q 端断开,M0.1 失电(M0.1 的线圈仅闭合了一个扫描周期)。

2) M0.1 常开触点闭合,Q0.0 线圈得电自锁,使接在 PLC 的 Q0.0 端的 KA 线圈吸合。

KA 线圈的吸合使主电路中的 KA 常开触点闭合,使接触器 KM 线圈通电吸合。KM 线圈的吸合使其主触点闭合,电动机启动。

(2) 立即停止:点动 SB_2,I0.1 常开触点先断开后闭合,使 Q0.0 线圈的自锁状态破坏,Q0.0 线圈失电。Q0.0 线圈的失电造成中间继电器 KA 线圈失电断开。KA 线圈的失电使其主电路中的常开触点由闭合到断开,从而使继电器 KM 线圈失电,其主触点由闭合变为断开状态,电动机停止运转。

当主电路由于过载造成热继电器 FR 动作时,PLC I0.2 端子的 FR 常闭触点断开,电动机停转。

操作步骤

其操作步骤分别为:

(1) 在 XK-SX2C 高级维修电工实训台上,使用跨接线完成如图 3-9 所示的 PLC 控制电路的接线和如图 3-10 所示的主电路接线。

(2) 使用万用表分别测量主电路和控制电路,检查电路是否有短路、断路或接错线。如果存在问题,仔细检查,直到排除故障。

(3) 闭合实训台总电源,开启电脑并打开博图软件。

(4) 在博图软件的编程界面中输入如图 3-11 所示程序,闭合 PLC 供电的 DC 24V 电源开关,将程序其下载到 PLC 中,设置 PLC 到运行状态并监控程序。

任务演示

其演示步骤分别为:

(1) 点动 SB_1,中间继电器 KA 线圈延时 10s 吸合。

(2) 点动 SB_2,KA 线圈立即断开。

(3) 闭合主电路隔离开关 QS。

(4) 再次点动 SB_1,KA 线圈延时 10s 吸合,同时 KM 线圈 KM 吸合,电动机启动。

(5) 点动 SB_2,KA 和 KM 的线圈同时立即断开。

(6) 演示结束后,将所有电源断开。

任务总结

总结整个任务实施过程,查找不足,尤其对于出现的故障要高度重视,分析总结后记录在表 3-3 中。

3.1.5 任务评价

对电动机延时启动主控电路的接线与调试任务的完成情况进行评分,并将评分结果填入表3-2内。

表3-2 评分表

任务内容	考核要求	评分标准	配分	扣分	得分
继电器控制主控电路接线与演示	正确完成继电器主电路及控制电路的接线、演示及说明	(1)电路不能实现延时启动扣10分; (2)电路不能实现立即停止扣5分; (3)不能对操作现象进行分析和说明的每处扣5分	30		
PLC控制系统主控电路接线	正确完成主电路及PLC控制电路的接线	正确完成主电路及PLC控制电路的接线,每接错一根扣10分	25		
PLC程序编写与下载	按照任务要求将PLC程序输入、下载与监控	(1)软件使用不熟练,不能完成程序的输入操作扣15分; (2)不会进行程序的IP地址设置与下载扣5分; (3)不能通过博图软件控制PLC的停止和运行状态扣5分; (4)不会操作博图软件监控程序的运行扣5分	15		
PLC控制系统运行演示	正确演示电动机的启动和停止,并能结合程序、硬件进行说明	(1)点动启动按钮后,不能正确说明电动机的延时启动过程扣10分; (2)点动停止按钮后,不能正确说明电动机的立即停止过程扣10分	20		
安全文明生产	遵守8S管理制度,遵守安全管理制度	(1)未穿戴劳动保护用品扣10分; (2)操作存在安全隐患,每次扣5分,扣完为止; (3)操作现场未及时整理整顿,每次发现扣2分,扣完为止	10		
总 分					

3.1.6 任务小结

任务实施过程记录单见表3-3。

表 3-3 任务实施过程记录单

	任务实施过程记录单	
故障 1	故障现象	
	故障原因	
	排除过程	
故障 2	故障现象	
	故障原因	
	排除过程	
故障 3	故障现象	
	故障原因	
	排除过程	
总　结		

3.1.7 任务拓展

任务要求：点动启动按钮 SB_1，电动机立即启动，运行 8s 后停止。若在电动机运行期间点动按钮 SB_2，电动机立即停止。请按照上述设计要求完成以下任务。

(1) 设计继电器控制系统的主控电路图并完成接线与调试。
(2) 完成 PLC 的 I/O 地址分配。
(3) 设计 PLC 控制系统的主控电路图，并完成接线。
(4) 编写 PLC 控制程序，完成程序的下载、运行与软硬件调试。

任务 3.2　电动机的延时停止主电路及控制电路的设计与实现

3.2.1 任务描述

本任务分别利用继电器系统和 PLC 控制系统完成对三相异步电动机的延时停止主控电路的接线与调试。

3.2.2 任务目标

(1) 掌握三相异步电动机的延时停止继电器控制系统的主控电路的设计、接线与调试。
(2) 理解西门子 S7-1200 PLC 置位、复位指令的工作原理。
(3) 理解西门子 S7-1200 PLC 上升沿、下降沿指令的工作原理。
(4) 掌握三相异步电动机的延时停止 PLC 控制系统的主控电路的接线。
(5) 理解 PLC 程序的设计思路并掌握程序的下载与调试。

3.2.3 任务相关知识

3.2.3.1　置位、复位指令

在位逻辑指令中，与置位和复位相关的指令共有三组 6 个指令，下面分别加以介绍。

置位输出和复位输出指令

S（置位输出）、R（复位输出）指令将指定的位操作数置位和复位。如果同一操作数的 S 线圈和 R 线圈同时断电，指定操作数的信号状态不变。置位输出指令与复位输出指令最主要的特点是有记忆和保持功能。

如图 3-12（a）所示，如果 I0.4 的常开触点闭合，Q0.5 变为 1 状态并保持该状态。即使 I0.4 的常开触点断开，Q0.5 也仍然保持 1 状态。I0.5 的常开触点闭合，Q0.5 变为 0 状态并保持该状态。即使 I0.5 的常开触点断开，Q0.5 也仍然保持 0 状态。程序对应时序图如图 3-12（b）所示。

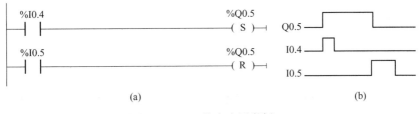

图 3-12　S/R 指令应用举例（1）

图 3-13 是将图 3-11 程序用置位输出和复位输出指令编程，从而实现同样的控制效果。I0.0 将 M0.0 置 1，使 M0.0 常开触点闭合，定时 IN 端接通开始计时。当计时时间达到定时器设定值时，定时器 Q 端接通，使 M0.1 线圈得电，M0.1 常开触点闭合。M0.1 常开触点闭合，使 M0.0 复位，M0.0 常开触点断开，定时器停止计时，另外，Q0.0 置 1，电动机启动运行。I0.1 和 I0.2 在程序中使用常闭形式是由于在 PLC 外部电路中，I0.1 和 I0.2 接入的是按钮和热继电器的常闭形式。

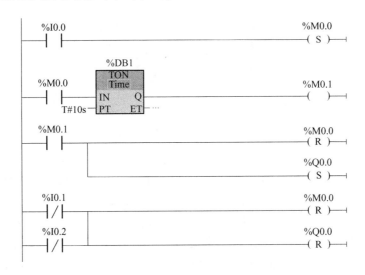

图 3-13　S/R 指令应用举例（2）

使用 S、R 指令时需要注意：
（1）在一个程序中，可以对同一个位进行多次的置位和复位操作。

（2）使用 S、R 指令时，建议不要使该逻辑行长时间接通，否则会降低程序的严谨性。

（3）在一个程序中，同一个线圈最好不要使用线圈输出和置位指令。

置位/复位触发器和复位/置位触发器

置位/复位触发器的梯形图形式如图 3-14（a）所示，其标识符是 SR。微复位/置位触发器的梯形图形式如图 3-14（b）所示，其标识符是 RS。

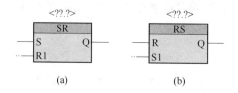

图 3-14　SR 和 RS 指令的梯形图形式

SR 指令是根据输入 S 和 R1 的信号状态，置位或复位指定操作数的位。如果输入 S 的信号接通，且输入 R1 的信号状态为断开，则将指定的操作数置 1；如果输入 S 的信号断开，且输入 R1 的信号接通，则将指定的操作数复位为 0。输入 R1 的优先级高于输入 S，输入 S 和 R1 的信号状态都接通时，指定操作数的信号状态将复位（复位优先）。如果两个输入 S 和 R1 的信号状态都断开，则不会执行该指令，因此操作数的信号状态保持不变。程序举例如图 3-15 所示，若 I0.0 和 I0.1 输入端分别接常开按钮 SB_1 和 SB_2，则点动 SB_1，M0.1 和 Q0.0 均置 1；点动 SB_2，M0.1 和 Q0.0 均复位；若 SB_1 和 SB_2 同时按下，则 M0.1 和 Q0.0 均复位。M0.1 和 Q0.0 均为 SR 指令的操作数，其中，线圈 Q0.0 为可选项。

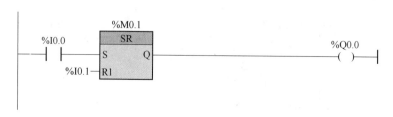

图 3-15　SR 指令应用举例

RS 指令根据输入 R 和 S1 的信号状态，复位或置位指定操作数的位。如果输入 R 的信号状态接通，且输入 S1 的信号状态断开，则指定的操作数将复位。如果输入 R 的信号状态断开，且输入 S1 的信号接通，则将指定的操作数置 1。输入 S1 的优先级高于输入 R，当输入 R 和 S1 的信号状态均接通时，将指定操作数的信号状态置位 1（置位优先）；如果两个输入 R 和 S1 的信号状态都断开时，则不会执行该指令，因此操作数的信号状态保持不变。程序举例如图 3-16 所示，若 I0.0 和 I0.1 输入端分别接常开按钮 SB_1 和 SB_2，则点动 SB_1，M0.0 和 Q0.0 均复位；点动 SB_2，M0.0 和 Q0.0 均置 1；若 SB_1 和 SB_2 同时按下，则 M0.0 和 Q0.0 均置 1。M0.0 和 Q0.0 均为 RS 指令的操作数，其中，线圈 Q0.0 为可选项。

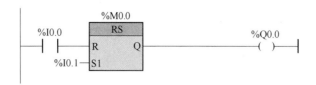

图 3-16 RS 指令应用举例

下面是 SR 指令的一个应用举例，设计要求如下：
(1) 若抢答器有三个输入按钮，分别接 I0.0、I0.1 和 I0.2。
(2) 输出有三个指示灯，分别接 Q4.0、Q4.1 和 Q4.2，复位输入按钮接 I0.4。

要求：三人中任意抢答，谁先按按钮，谁的指示灯优先亮，且只能亮一盏灯。进行下一问题时主持人按复位按钮，抢答重新开始。

PLC 程序如图 3-17 所示。假设先点动 I0.0 所连接按钮，则 Q0.4 得电，其对应的指示灯亮，程序中 Q0.4 的常闭触点由闭合变为断开。这样，点动 I0.1 和 I0.2 所对应按钮都不会执行它们所串接 SR 指令。

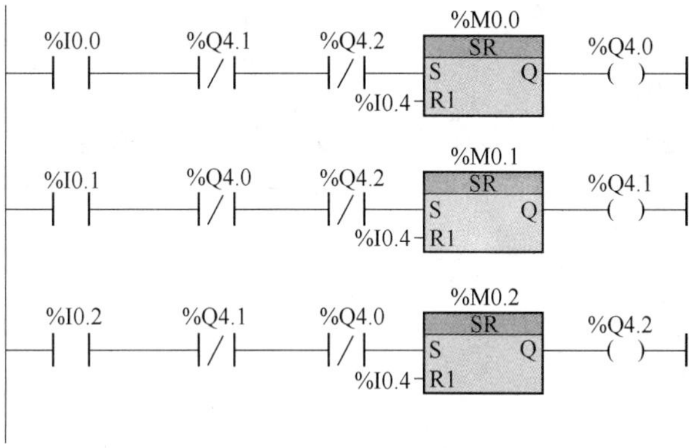

图 3-17 抢答器控制 PLC 参考程序

置位位域指令和复位区域指令

置位位域指令的标识符是 SET_BF，该指令可对从某个特定地址开始的多个位进行置位。复位区域指令的标识符是 RESET_BF，该指令可对从某个特定地址开始的多个位进行复位。如图 3-18 所示，I0.0 的通断，使 Q0.0 开始的 5 个输出线圈置 1，即 Q0.0~Q0.4 被置 1。I0.1 的通断，将从 Q0.2 开始的 2 个位被复位，最后的运行结果是仅有 Q0.0、Q0.1、Q0.4 被置 1。

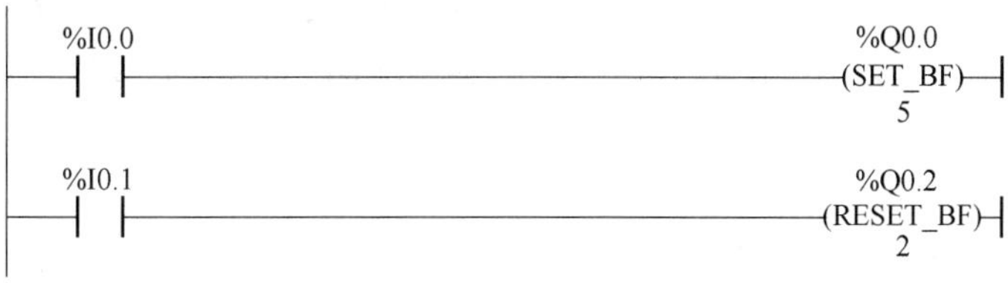

图 3-18 SET_BF 和 RESET_BF 应用举例

练一练

分别使用 SR 和 RS 指令完成如图 3-11 所示的程序设计。

3.2.3.2 上升沿指令和下降沿指令

当信号状态变化时就产生跳变沿。当从 0 变到 1 时，产生一个上升沿（或正跳变）；若从 1 变到 0，则产生一盒下降沿（或负跳变）。如图 3-19 所示，当与 PLC 的 I0.0 所连接的常开按钮 SB 被点动时，I0.0 状态从 0 变到 1 会产生一个上升沿，从 1 变到 0 会产生一个下降沿。

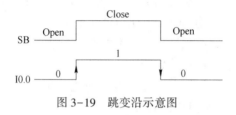

图 3-19 跳变沿示意图

在 S7-1200 PLC 中，与跳变沿检测有关的指令共有四组（8 个），这里仅介绍两组指令。

操作数信号扫描指令

操作数信号扫描包括扫描操作数的信号上升沿指令和扫描操作数的信号下降沿指令。该组指令用来检测操作数的上升沿信号和下降沿信号，同时使逻辑行接通一个扫描周期，如图 3-20 所示。

图 3-20 操作数信号扫描指令应用举例（1）

中间有 P 的触点是扫描操作数的信号上升沿指令，在 I0.0 的上升沿，该触点接通一个扫描周期将 Q0.0~Q0.3 置 1，M4.3 为边沿存储位，用来存储上一次扫描循环时 I0.0 的状态。通过比较 I0.0 前后两次循环的状态，来检测信号的边沿。边沿存储位的地址只能在程序中使用一次，不能用代码块的临时局部数据或 I/O 变量作为边沿存储位。中间有 N 的触点是扫描操作数的信号下降沿指令，在 I0.1 的下降沿，RESET_BF 的线圈"通电"一个扫描周期，将 Q0.2 和 Q0.3 复位，该触点下面的 M4.4 为边沿存储位。

如图 3-21 所示，该程序当 I0.0 所连接常开按钮闭合时，Q0.0 立即得电；当 I0.0 所连接常开按钮断开时，Q0.0 延时 5s 失电。

扫描 RLO 的信号边沿指令

在西门子 PLC 中，状态字的位 1 称为逻辑操作结果 RLO（Result of Logic Operation），该位存储逻辑指令或算术比较指令的结果。在逻辑行中，RLO 位的状态能够表示有关信号

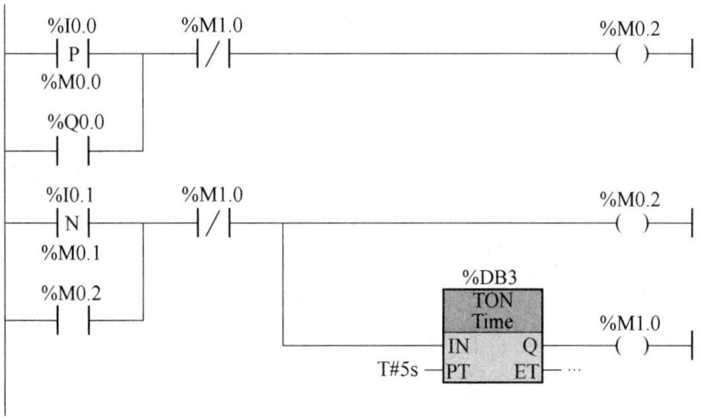

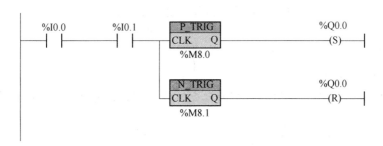

图 3-21 操作数信号扫描指令应用举例（2）

流的信息。RLO 的状态为 1，则表示有信号流（通）；为 0，则表示无信号流（断）。可用 RLO 触发跳转指令。

在 S7-1200 PLC 中，扫描 RLO 的信号边沿的指令分别是扫描 RLO 的信号上升沿指令（P_TRIG）和扫描 RLO 的信号下降沿指令（N_TRIG）。如图 3-22 所示，在流进 P_TRIG 指令的 CLK 输入端的能流（即 RLO）由断开到接通时（上升沿），P_TRIG 指令的 Q 端输出脉冲宽度为一个扫描周期的能流（接通一个扫描周期），将 Q0.0 置 1，指令下方的 M8.0 是脉冲存储位。在流进 N_TRIG 指令的 CLK 输入端的能流由接通到断开时（下降沿），N_TRIG 指令的 Q 端输出一个扫描周期的能流（接通一个扫描周期），将 Q0.0 复位，指令下方的 M8.1 是脉冲存储器位。P_TRIG 指令与 N_TRIG 指令不能放在电路的开始处和结束处。

图 3-22 P_TRIG 指令和 N_TRIG 指令应用举例

练一练

用 P_TRIG 指令和 N_TRIG 指令实现图 3-21 程序的相同功能。

3.2.4 任务实施

3.2.4.1 电动机的延时停止主控电路的接线与调试（继电器控制系统）

任务要求

点动启动按钮，电动机立即启动；点动停止按钮，电动机延时 10s 停止；点动急停按钮，电动机立即停止。完成主电路和控制电路的接线及调试，电路应具备必要的短路和过载保护措施。

完成后对本项目进行评分，详见表3-5。

任务分析

按下启动按钮，控制电动机的接触器应立即吸合，使其主触点启动电动机。要使按下停止按钮，电动机延时停止，就必须用该动作启动一个时间继电器通电计时。为保证时间继电器通电时间能到10s，需要借助中间继电器的自锁来实现。定时时间到后，使时间继电器的常闭延时触点断开，使控制电动机工作的接触器线圈失电，电动机停止。主电路应有急停按钮、熔断器和热继电器；控制电路应有熔断器，并有热继电器的常闭触点。

综上所述，使用时间继电器实现电动机延时启动主控电路如图3-23所示。

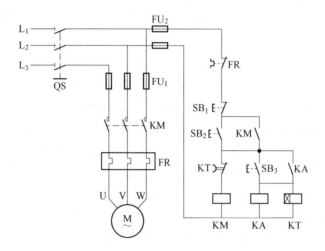

图3-23 电动机延时停止继电器主控电路图

操作步骤

其操作步骤分别为：

（1）在XK-SX2C高级维修电工实训台上，使用跨接线完成如图3-23所示的主电路和控制电路连接。

（2）设置时间继电器的动作时间为10s。

（3）使用万用表分别测量主电路和控制电路，检查电路是否有短路、断路或接错线。如果存在问题，仔细检查，直到排除故障。

任务演示

其演示步骤分别为：

（1）闭合实训台总电源和电路，隔离开关QS。

（2）点动SB_2，电动机立即启动。

（3）点动SB_3，中间继电器线圈吸合，时间继电器指示灯亮，10s后接触器线圈断电释放，电动机停止，同时中间继电器和时间继电器断电。

（4）再次点动SB_2，电动机立即启动，点动SB_1，电动机立即停止。

（5）断开隔离开关QS和实训台总电源。

3.2.4.2 电动机的延时停止主控电路的接线与调试（PLC控制系统）

任务要求

点动与PLC输入点连接的启动按钮，电动机立即启动；点动与PLC输入点连接停止按钮，电动机延时10s停止；电动机运行状态下点动急停按钮，电动机立即停止。完成主电路和PLC控制电路的接线及调试，电路应具备必要的短路和过载保护措施。

完成后对本项目进行评分，详见表3-5。

任务分析

设计PLC控制系统，首先应根据设计要求进行分析，确定PLC的I/O地址分配，电动机延时停止PLC控制系统I/O地址分配表见表3-4。

表3-4 电动机延时停止PLC控制系统I/O地址分配表

输入信号				输出信号			
序号	功能	元件	地址	序号	控制对象	元件	地址
1	启动按钮	SB_1	I0.0	1	中间继电器	KA	Q0.0
2	停止按钮	SB_2	I0.1	—	—	—	—
3	急停按钮	SB_3	I0.2	—	—	—	—
4	过载保护	FR	I0.3	—	—	—	—

在PLC的地址分配表基础上，设计PLC的外部接线图，如图3-24所示。

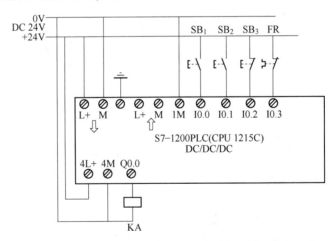

图3-24 电动机延时停止PLC控制电路接线图

在主电路中，对接触器的控制需要通过将中间继电器的常开触点来控制接触器线圈，如图3-25所示。

对照图3-25和图3-10可知，这两个电路图完全一样，这也是PLC控制系统相对于继电器控制系统的优点之一，即：PLC外部接线图以及主电路不修改或部分修改，就可实现控制功能的改变。

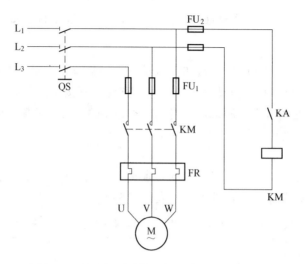

图 3-25 电动机延时停止主电路接线图（PLC）

在 PLC 主电路和控制电路设计完成的基础上进行 PLC 程序的设计，本项目给出了三种设计方法，参考程序分别如图 3-26~图 3-28 所示。

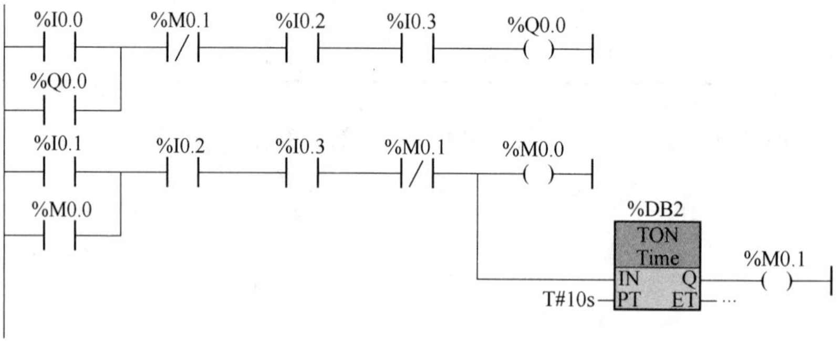

图 3-26 电动机延时停止 PLC 参考程序（1）

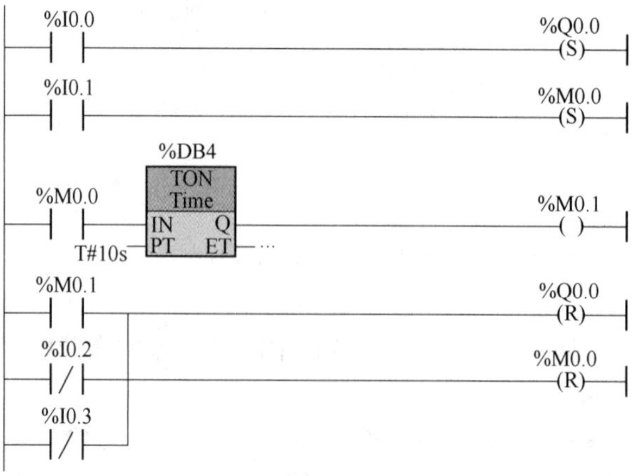

图 3-27 电动机延时停止 PLC 参考程序（2）

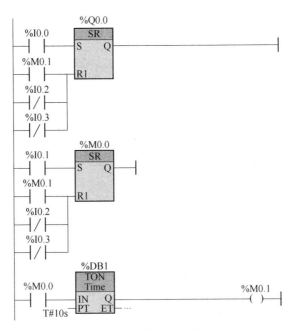

图 3-28 电动机延时停止 PLC 参考程序（3）

以图 3-27 所示程序，讲解 PLC 程序执行及硬件设备运行：

（1）立即启动：点动 SB_1，Q0.0 置 1，KA 线圈吸合。KA 线圈的吸合使主电路中的 KA 常开触点闭合，接触器 KM 线圈通电吸合。KM 线圈的吸合使其主触点闭合，电动机启动。

（2）延时停止：点动 SB_2，I0.1 接通后断开，使 M0.0 线圈置 1，M0.0 线圈置 1 使其常开触点闭合，定时器开始计时。当定时器计时到 10s 时，定时器 Q 端接通，使 M0.1 线圈接通。由于 M0.1 线圈接通，将 Q0.0 复位，KA 线圈断电释放。KA 线圈的释放使主电路中的 KA 常开触点断开，接触器 KM 线圈通电断电释放。KM 线圈的释放使其主触点断开，电动机停止。

（3）立即停止。点动 SB_3，I0.2 常闭触点接通后断开，将 Q0.0 和 M0.0 均复位，电动机停止运转。当主电路由于过载造成热继电器 FR 动作，使接下 PLC I0.3 端子的 FR 常闭触点断开，电动机停转。

操作步骤

其操作步骤分别为：

（1）在 XK-SX2C 高级维修电工实训台上，使用跨接线完成如图 3-24 所示的 PLC 控制电路的接线和如图 3-25 所示的主电路接线。

（2）使用万用表分别测量主电路和控制电路，检查电路是否有短路、断路或接错线。如果存在问题，仔细检查，直到排除故障。

（3）闭合实训台总电源，开启电脑并打开博图软件。

（4）在博图软件的编程界面中输入如图 3-26 所示的程序，闭合 PLC 供电的 DC24V 电源开关，将程序下载到 PLC 中，将 PLC 设置到运行状态并监控程序。

(5) 通过点动 SB_1、SB_2 和 SB_3，观察线圈的状态、电动机的运行情况以及程序的运行情况。

(6) 分别将如图 3-27 和图 3-28 所示的 PLC 程序下载和监控，通过操作(5)观察程序的运行过程。

任务演示

其演示步骤分别为：

(1) 点动 SB_1，中间继电器线圈 KA 立即吸合。

(2) 点动 SB_2，KA 线圈延时 10s 断电释放。

(3) 再次点动 SB_1，KA 立即吸合。

(4) 点动 SB_3，KA 立即断电释放。

(5) 闭合主电路隔离开关 QS。

(6) 点动 SB_1，中间继电器线圈 KA 和接触器线圈 KM 立即吸合，电动机立即启动。

(7) 点动 SB_2，KA 和 KM 线圈延时 10s 断电释放，电动机停止。

(8) 再次点动 SB_1，电动机立即启动。

(9) 点动 SB_3，电动机立即停止。

(10) 演示结束后，将所有电源断开。

任务总结

总结整个任务实施过程，查找不足，尤其对于出现的故障要高度重视，分析总结后记录在表 3-6 中。

3.2.5 任务评价

对电动机延时停止主控电路的接线与调试任务的完成情况进行评分，并将评分结果填入表 3-5 内。

表 3-5 评分表

任务内容	考核要求	评分标准	配分	扣分	得分
继电器控制主控电路接线与演示	正确完成继电器主电路及控制电路的接线、演示及现象说明	(1) 电路不能实现立即启动扣除 5 分； (2) 电路不能实现延时停止扣除 10 分； (3) 电路不能实现急停功能扣 5 分； (4) 不能对操作现象进行分析和说明的每处扣 5 分	30		
PLC 控制主控电路接线	正确完成主电路及 PLC 控制电路接线	主电路及 PLC 控制电路的接线，每接错一根扣 10 分	20		

续表 3-5

任务内容	考核要求	评分标准	配分	扣分	得分
PLC 程序编写与下载	按照任务要求将 PLC 程序输入、下载与监控	（1）软件使用不熟练，不能完成程序的输入操作扣 15 分； （2）不会进行程序的 IP 地址设置与下载扣 5 分； （3）不能通过博图软件控制 PLC 的停止和运行状态扣 5 分； （4）不会操作博图软件监控程序的运行扣 5 分	20		
PLC 控制系统运行演示	正确演示电动机的启动和停止，并能结合程序、硬件进行说明	（1）点动启动按钮后，不能正确说明电动机的立即启动过程扣 5 分； （2）点动停止按钮后，不能正确说明电动机的延时停止过程扣 10 分； （3）点动急停按钮，不能正确说明电动机的立即停止过程扣 5 分	20		
安全文明生产	遵守 8S 管理制度，遵守安全管理制度	（1）未穿戴劳动保护用品扣 10 分； （2）操作存在安全隐患，每次扣 5 分，扣完为止； （3）操作现场未及时整理整顿，每次发现扣 2 分，扣完为止	10		
总　分					

3.2.6 任务小结

任务实施过程记录单见表 3-6。

表 3-6　任务实施过程记录单

任务实施过程记录单		
故障 1	故障现象	
	故障原因	
	排除过程	
故障 2	故障现象	
	故障原因	
	排除过程	
故障 3	故障现象	
	故障原因	
	排除过程	
总　结		

3.2.7 任务拓展

任务要求：点动启动按钮 SB_1，电动机延时 5s 立即启动；点动延时停止按钮 SB_2，电动机延时 8s 停止。若在电动机运行期间点动急停按钮 SB_3，电动机立即停止。请按照上述设计要求完成以下任务：

(1) 设计继电器控制系统的主控电路图并完成接线与调试。
(2) 确定 PLC 的 I/O 地址分配。
(3) 设计 PLC 控制系统的主控电路图，并完成接线。
(4) 用两种方法编写 PLC 控制程序，完成程序的下载、运行与软硬件调试。

任务 3.3 电动机的循环启停主电路及控制电路的设计与实现

3.3.1 任务描述

本任务分别利用继电器系统和 PLC 控制系统完成对三相异步电动机循环启停控制的主控电路的接线与调试。

3.3.2 任务目标

(1) 掌握三相异步电动机的循环启停控制的继电器控制系统主控电路的接线与调试。
(2) 理解西门子 S7-1200 PLC 比较指令的工作原理。
(3) 理解西门子 S7-1200 PLC 计数器指令的工作原理。
(4) 掌握三相异步电动机的循环启停控制的 PLC 控制系统主控电路的接线。
(5) 理解 PLC 程序的设计思路并掌握程序的下载与调试。

3.3.3 任务相关知识

3.3.3.1 比较指令

在 S7-1200 PLC 中，与比较有关的指令有两个操作数的等于比较、不等于比较、大于比较、小于比较、范围比较、变量比较等，这里仅讲解常用的六个比较指令，其梯形图形式及接通条件见表 3-7。

表 3-7 常用比较指令梯形图形式及接通条件

指令	逻辑行接通条件	指令	逻辑行接通条件
─┤ IN1 == ??? IN2 ├─	IN1 = IN2	─┤ IN1 >= ??? IN2 ├─	IN1 ≥ IN2
─┤ IN1 <> ??? IN2 ├─	IN1 ≠ IN2	─┤ IN1 < ??? IN2 ├─	IN1 < IN2
─┤ IN1 > ??? IN2 ├─	IN1 > IN2	─┤ IN1 <= ??? IN2 ├─	IN1 ≤ IN2

在 PLC 基本程序设计中，定时器和计数器的比较较为常用，可以是两个定时器（或计数器）的当前值比较，也可以是一个定时器（或计数器）的当前值与常数的比较。适当的采用比较指令编程可以简化程序。如图 3-29 所示，若 I0.0 连接按钮常开 SB_1，I0.1 连接按钮常开 SB_2，则当点动 SB_1 时，Q0.0 延时 1s 得电，Q0.0 得电 2s 后 Q0.1 得电，Q0.1 得电 2s 后 Q0.2 得电；点动 SB_2，Q0.2 立即失电、1s 后 Q0.1 失电，Q0.1 失电 1s 后 Q0.0 失电。在程序中，给 5s 定时器命名为 T1，T1 定时器的当前值用"T1.ET"表示，2s 定时器的当前值用"MD10"表示，这两种方法均能读取定时器的当前值。

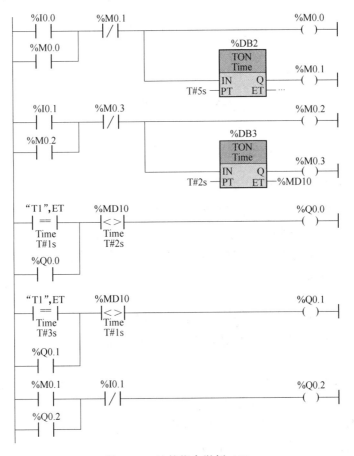

图 3-29　比较指令举例（1）

有时需要一个或几个输出点循环得电与失电，如果在编程中使用比较指令也可使程序编写得到简化，如图 3-30 所示程序。若 I0.0 连接按钮常开 SB_1，I0.1 连接按钮常开 SB_2，Q0.0 和 Q0.1 各连接一盏指示灯 HL_1 和 HL_2。当点动 SB_1 时，HL_1 立即亮，5s 后熄灭，再过 2s HL_2 亮，HL_2 亮 4s 后熄灭，再过 1s，HL_1 再次亮，如此循环。当点动 SB_2 时，HL_1 和 HL_2 立即熄灭。在程序中，定时器 T1（12s 定时器）前面逻辑行接 M0.1 常闭触点是为了实现定时器的当前值在 0~12s 间进行循环。

练一练

（1）请将如图 3-29 所示程序用置位和复位指令编程实现。

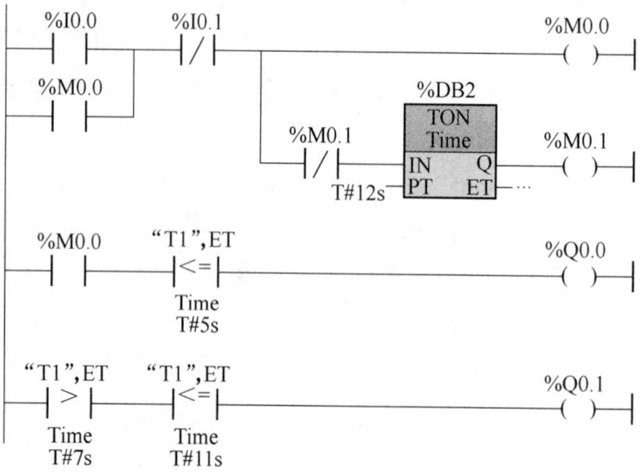

图 3-30 比较指令举例（2）

（2）请将如图 3-30 所示程序用比较指令的等于指令和不等于指令编程实现。

3.3.3.2 计数器指令

S7-1200 有 3 种计数器，其分别为加计数器（CTU）、减计数器（CTD）和加减计数器（CTUD）。它们属于软件计数器，其最大计数速率受到所在的 OB 的执行速率的限制。如果需要速率更高的计数器，可以使用 CPU 内置的高速计数器。调用计数器指令时，需要生成保存计数器数据的背景数据块。

加计数器指令

加计数器的标识符是 CTU。该指令共有 5 个参数：

（1）计数输入 CU：当 CU 端的值从 0 变为 1 时（信号上升沿），CTU 当前计数值加 1，最大到 32767。

（2）复位输入 R：当 R 端的值从 0 变为 1 时（信号上升沿），当前 CTU 计数值归 0。

（3）设定值输入 PV：计数器设定值输入端。

（4）计数器状态输出 Q：输出 Q 的信号状态由参数 PV 决定。如果当前计数器值大于或等于参数 PV 的值，则将输出 Q 的信号状态置 1（接通）。

（5）当前计数器值 CV：存储计数器的当前计数值。

加计数器指令工作原理如图 3-31 所示。其中，图 3-31（a）是 PLC 程序；图 3-31（b）是程序执行时序图。

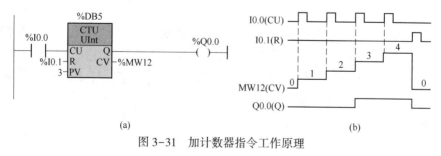

图 3-31 加计数器指令工作原理

减计数器指令

减计数器的标识符是 CTD。该指令共有五个参数：

(1) 计数输入 CD：当 CU 端的值从 0 变为 1 时（信号上升沿），CTD 当前计数值减 1，最小到 32768。

(2) 装载输入 LD：当 LD 端的值从 0 变为 1 时（信号上升沿），CTD 装载计数值设定值 PV，计数器当前值为 PV 所设定的值。

(3) 设定值输入 PV：计数器设定值输入端。

(4) 当前计数器值 CV：存储计数器的当前计数值。

(5) 计数器状态输出 Q：当参数 CV 的值等于或小于 0，则计数器将输出 Q 的信号状态置 1（接通）。

减计数器指令工作原理如图 3-32 所示。其中图 3-32 (a) 是 PLC 程序；图 3-32 (b) 是程序执行时序图。

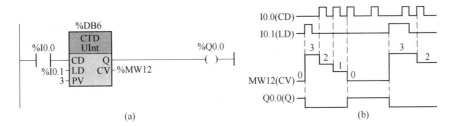

图 3-32 减计数器指令工作原理

加减计数器指令

加减计数器的标识符是 CTUD，该指令是加计数器和减计数器的结合，共有八个参数。其中，QU 是加计数状态输出，QD 是减计数状态输出。其他参数说明参考加计数器和减计数器的参数说明。加减计数器指令工作原理如图 3-33 所示，其中图 3-33 (a) 是 PLC 程序；图 3-33 (b) 是程序执行时序图。

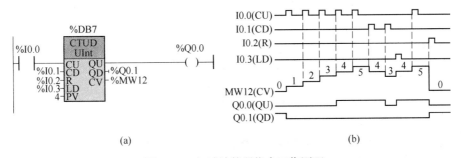

图 3-33 加减计数器指令工作原理

计数器指令的应用较为普遍，如流水线上工件的计数、电动机循环运行的次数计数等。如图 3-34 所示，PLC 程序实现了用一个按钮对 Q0.0 进行得电与失电的控制。其中：加计数器的 Q 端接 M0.0 线圈的作用是计数器计数到设定值 2 后，由于 M0.0 线圈的得电使其常开触点闭合，复位计数器，从而实现操作的可重复性。其中，"IEC_Counter_0_DB". CV 是计数器的当前值。

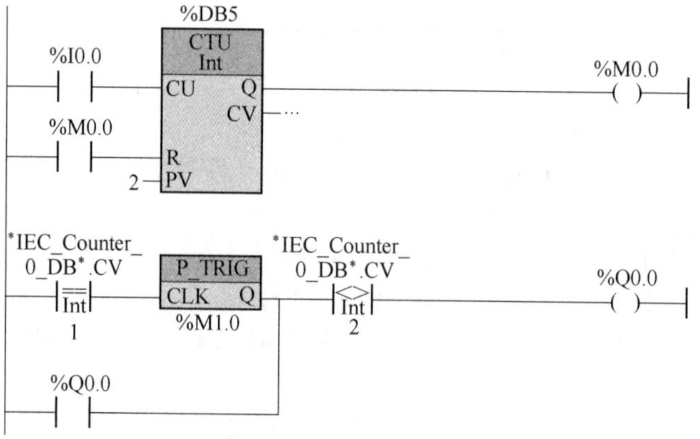

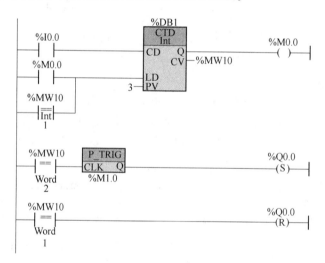

图 3-34 计数器举例（1）

图 3-35 是用另一种编程方法实现该功能。在该程序中，若减计数器当前值为 0，则在 PLC 运行开始时，减计数器 Q 端就会接通，从而使 M0.0 线圈得电。其常开触点闭合，将设定值 3 装载，计数器当前值为 3。当 I0.0 所对应按钮被点动两次时，计数器当前值为 1，LD 端的比较指令接通，重新将减计数器设定值 3 装载。

图 3-35 计数器举例（2）

对于相对复杂的设计要求，在编程中可能需要使用多个计数器。投票箱示意图如图 3-36 所示，该投票箱上有 $SB_1 \sim SB_4$ 四个按钮（对应 PLC 的 I0.0~I0.3），分别对应选甲、选乙、弃权和复位；三个指示灯 $HL_1 \sim HL_3$（对应 PLC 的 Q0.0~Q0.2），分别对应甲当选、乙当选和票数相等。每次投票前都要先点动复位按钮将计数器归零。一个班有 10 名学生，甲、乙两人竞选班长，甲、乙两人也参加投票，只有投票结束后，才显示投票结果，谁的票数多谁当选，对应的指示

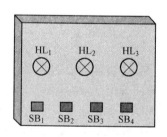

图 3-36 投票箱示意图

·230·

灯亮。

这里采用两种方法进行编程设计，如图 3-37 所示。在该程序中，采用 3 个加计数器实现设计要求。DB2 对应计数器计算甲的得票数；DB3 对应计数器计算乙的得票数；DB4 对应计数器计算总投票数。当投票结束后，在程序段最后，MW14 的值等于 10，该比较指令接通，从而使甲乙得票数的比较反应在对应指示灯的亮灭状态。三个计数器的设定值可以在设定范围内任意设置，在程序中仅使用了对其当前值的读取，而没有使用定时器的输出功能。

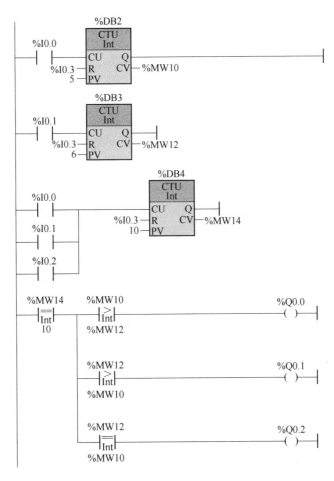

图 3-37 投票箱 PLC 程序设计（1）

还可以用 1 个加计数器和 1 个加减计数器实现设计要求，如图 3-38 所示。在这种设计方法中，加计数器计数总票数，加减计数器对甲乙的得票数进行加减计数。由程序中可以看出，由于首先点动复位按钮 SB_4 计数器进行了复位，加减计数器的当前值为 0。投票开始后，若甲得 1 票，则加减计数器加 1；若乙得 1 票，则加减计数器减 1。当投票结束后，可以通过将计数器的当前值与 0 比较，获知甲乙谁的票数多。

练一练

用减计数器和加计数器编写投票箱的 PLC 程序。

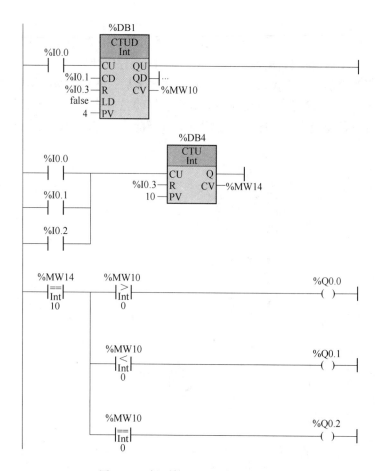

图 3-38 投票箱 PLC 程序设计（2）

3.3.4 任务实施

3.3.4.1 电动机循环启停主控电路的接线与调试（继电器控制系统）

任务要求

点动启动按钮，电动机立即启动，运行 10s 停止 5s，循环运行；点动急停按钮，电动机立即停止。完成主电路和控制电路的接线及调试，电路应具备必要的短路和过载保护措施。

完成后对本项目进行评分，详见表 3-9。

任务分析

确定电路的启动按钮为 SB_1，停止按钮为 SB_2，电动机运行状态对应时间继电器为 KT_1，电动机停止状态对应时间继电器为 KT_2。

按下启动按钮 SB_1，控制电动机的接触器线圈 KM 和 KT_1 线圈应立即吸合，使电动机启动并开始计时；电动机运行时间到 10s 时，KT_1 的延时断触点应使 KM 线圈和 KT_1 线圈断电释放，同时 KT_1 的延时闭合触点应使 KT_2 线圈吸合，为保证 KT_2 线圈吸合时间能到

5s，需要借助中间继电器 KA 的自锁来实现。当 KT_2 计时到 5s 时，其延时断触点将其自身自锁状态破坏，同时其延时闭合触点使接触器线圈 KM 和 KT_1 线圈再次吸合。

综上所述，使用时间继电器实现电动机延时启动主控电路如图 3-39 所示。

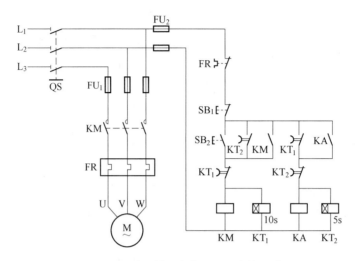

图 3-39 电动机循环启停继电器主控电路图

操作步骤

其操作步骤分别为：

（1）在 XK-SX2C 高级维修电工实训台上，使用跨接线完成如图 3-39 所示的主电路和控制电路连接。

（2）设置时间继电器 KT_1 的动作时间为 10s，KT_2 的动作时间为 5s。

（3）使用万用表分别测量主电路和控制电路，检查电路是否有短路、断路或接错线。如果存在问题，仔细检查，直到排除故障。

任务演示

其演示步骤分别为：

（1）闭合实训台总电源和电路，隔离开关 QS。

（2）点动 SB_2，电动机立即启动。

（3）电动机运行 10s 后停止，停止 5s 后又启动，循环运行。

（4）点动 SB_1，电动机立即停止。

（5）断开隔离开关 QS 和实训台总电源。

3.3.4.2　电动机的循环启停主控电路的接线与调试（PLC 控制系统）

任务要求

用同一个按钮实现电动机的启动与停止。第一次点动该按钮时，电动机立即启动，运行 10s 后停止 5s，然后再运行 10s 停止 5s，如此循环；第二次点动该按钮时，电动机立即停止。电动机运行状态下点动急停按钮，电动机立即停止。完成主电路和 PLC 控制电路的接线及调试，电路应具备必要的短路和过载保护措施。

完成后对本项目进行评分,详见表 3-9。

任务分析

设计 PLC 控制系统,首先应根据设计要求进行分析,确定 PLC 的 I/O 地址分配,电动机循环启停 PLC 控制系统 I/O 地址分配表见表 3-8。

表 3-8 电动机循环启停 PLC 控制系统 I/O 地址分配表

输入信号				输出信号			
序号	功能	元件	地址	序号	控制对象	元件	地址
1	启动/停止按钮	SB_1	I0.0	1	中间继电器	KA	Q0.0
2	急停按钮	SB_2	I0.1	—	—	—	—
3	过载保护	FR	I0.2	—	—	—	—

在 PLC 的地址分配表基础上,设计 PLC 的外部接线图如图 3-40 所示。

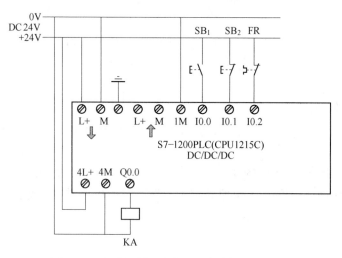

图 3-40 电动机循环启停 PLC 控制电路接线图

在主电路中,对接触器的控制需要通过将中间继电器的常开触点来控制接触器线圈。

在 PLC 主电路和控制电路设计完成的基础上进行 PLC 程序的设计,本项目给出了两种设计方法,第一种方法没有采用比较指令进行编程设计,参考程序如图 3-41 所示。程序中计数器前面的 M0.0 常开触点是为了使 SB_1 按钮操作具有重复性,即可以多次对电动机启动和停止操作,当然这里也可用比较指令,将计数器当前值和设定值 2 进行比较,将计数器复位。在程序段 2 中,使用 P_TRIG 指令来检测计数器当前值为数值 1 产生上升沿,其目的是为了使 Q0.0 自锁的条件接通一个瞬间,而不是一直接通(如果不第二次点动 SB_1 的情况下)。

如果使用比较指令来截取定时器的时间段,程序将相对简单,如图 3-42 所示。该设计通过在定时器前增加 M0.2 常闭触点,使定时器的当前值在 0~15s 间循环,通过用比较指令截取前 10s,从而使电动机循环启停。

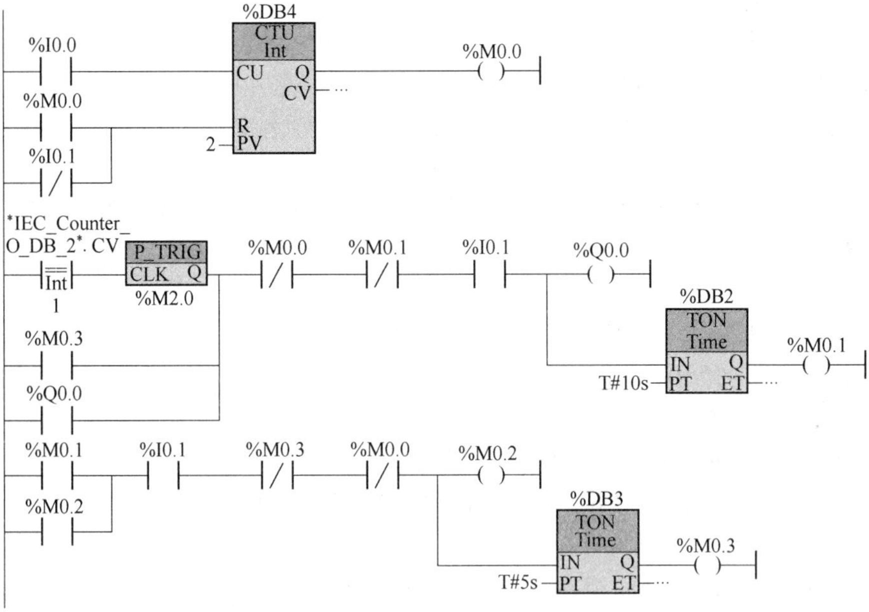

图 3-41 电动机循环启停 PLC 参考程序（1）

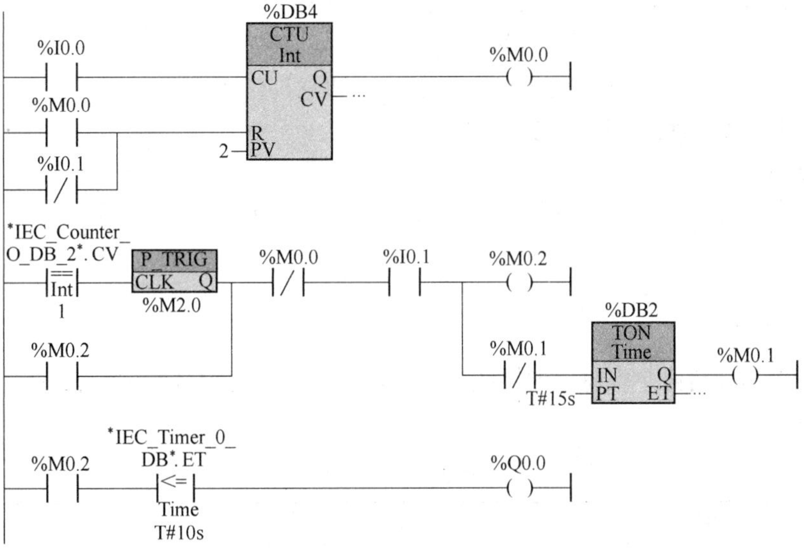

图 3-42 电动机循环启停 PLC 参考程序（2）

操作步骤

其操作步骤分别为：

（1）在 XK-SX2C 高级维修电工实训台上，使用跨接线完成如图 3-40 所示的 PLC 控制电路的接线和如图 3-42 所示的主电路接线。

（2）使用万用表分别测量主电路和控制电路，检查电路是否有短路、断路或接错线。如果存在问题，仔细检查，直到排除故障。

(3) 闭合实训台总电源，开启电脑并打开博图软件。

(4) 在博图软件的编程界面中输入如图 3-41 所示的程序，闭合 PLC 供电的 DC 24V 电源开关，将程序下载到 PLC 中，将 PLC 设置到运行状态并监控程序。

(5) 通过点动 SB_1 和 SB_2，观察线圈的状态、电动机的运行情况以及程序的运行情况。

(6) 再将如图 3-42 所示的 PLC 程序下载和监控，通过操作(5)观察程序的运行过程。

任务演示

其演示步骤分别为：

(1) 首次点动 SB_1，中间继电器线圈 KA 立即吸合。

(2) KA 线圈延时 10s 断电释放。

(3) 再过 5s，KA 线圈再次吸合。

(4) 再次点动 SB_1，KA 立即断电释放。

(5) 在 KA 线圈吸合状态下，点动 SB_2，KA 线圈立即断电释放。

(6) 闭合主电路隔离开关 QS。

(7) 首次点动 SB_1，中间继电器线圈 KA 和接触器线圈 KM 立即吸合，电动机立即启动。

(8) 10s 后 KA 线圈和 KM 线圈同时断电释放，电动机停止。

(9) 电动机停止 5s 后再次启动，在操作(6)和操作(7)循环。

(10) 再次点动 SB_1，电动机立即停止。

(11) 在电动机运行状态下，点动 SB_2，电动机立即停止。

(12) 演示结束后，将所有电源断开。

任务总结

总结整个任务实施过程，查找不足，尤其对于出现的故障要高度重视，分析总结后记录在表 3-10 中。

3.3.5 任务评价

对电动机循环启停主控电路的接线与调试任务的完成情况进行评分，并将评分结果填入表 3-9 内。

表 3-9 评分表

任务内容	考核要求	评分标准	配分	扣分	得分
继电器控制主控电路接线与演示	正确完成继电器主电路及控制电路的接线、演示及现象说明	(1) 电路不能实现立即启动扣 5 分； (2) 电路不能实现运行 10s 停止 5s 的循环扣 20 分； (3) 电路不能实现急停功能扣 5 分； (4) 不能对操作现象进行分析和说明的每处扣 5 分	30		

续表 3-9

任务内容	考核要求	评分标准	配分	扣分	得分
PLC 控制系统主控电路接线	正确完成主电路及 PLC 控制电路的接线	主电路及 PLC 控制电路的接线，每接错一根扣 5 分，该项扣完为止	15		
PLC 程序编写与下载	按照任务要求将 PLC 程序输入、下载与监控	（1）软件使用不熟练，不能完成程序的输入操作扣 15 分； （2）不会进行程序的 IP 地址设置与下载扣 5 分； （3）不能通过博图软件控制 PLC 的停止和运行状态扣 5 分； （4）不会操作博图软件监控程序的运行扣 5 分	25		
PLC 控制系统运行演示	正确演示电动机的启动和停止，并能结合程序、硬件进行说明	（1）首次点动按钮 SB_1，不能正确说明电动机的立即启动过程扣 5 分； （2）不能结合软硬件解释电动机循环启停的工作过程扣 10 分； （3）点动急停按钮，不能正确说明电动机的立即停止过程扣 5 分	20		
安全文明生产	遵守 8S 管理制度，遵守安全管理制度	（1）未穿戴劳动保护用品扣 10 分； （2）操作存在安全隐患，每次扣 5 分，扣完为止； （3）操作现场未及时整理整顿，每次发现扣 2 分，扣完为止	10		
总 分					

3.3.6 任务小结

任务实施过程记录单见表 3-10。

表 3-10 任务实施过程记录单

任务实施过程记录单		
故障 1	故障现象	
	故障原因	
	排除过程	
故障 2	故障现象	
	故障原因	
	排除过程	
故障 3	故障现象	
	故障原因	
	排除过程	
总 结		

3.3.7 任务拓展

3.3.7.1 两台电动机顺序启动与逆序停止主控电路设计、接线与调试

任务要求：首次点动 SB_1，电动机 M_1 立即启动，5s 后电动机 M_2 启动；第二次点动 SB_1，M_2 立即停止，8s 后电动机 M_1 停止。任何时候点动 SB_2，M_1、M_2 均立即停止。

根据任务要求完成以下任务：
(1) 设计继电器控制系统的主控电路图并完成接线与调试。
(2) 确定 PLC 的 I/O 地址分配。
(3) 设计 PLC 控制系统的主控电路图，并完成接线。
(4) 用两种方法编写 PLC 控制程序，完成程序的下载、运行与软硬件调试。

3.3.7.2 电动机循环启停 PLC 主控电路设计、接线与调试

任务要求：点动 SB_1，电动机延时 3s 启动，运行 5s 后停止，停止 2s 循环运行。循环 3 次后自动停止。任何时候点动 SB_2，电动机立即停止。

根据任务要求完成以下任务：
(1) 确定 PLC 的 I/O 地址分配。
(2) 设计 PLC 控制系统的主控电路图，并完成接线。
(3) 编写 PLC 控制程序，完成程序的下载、运行与软硬件调试。

3.3.8 项目小结

通过项目 3 中典型任务的学习和实践，理解了定时器、计数器、置/复位指令、上升沿指令和下降沿指令等 S7-1200 PLC 常用指令的工作原理，熟悉了电动机延时控制以及循环运行控制的继电器控制系统和 PLC 控制系统的硬件系统设计思路，掌握了 PLC 程序的设计方法。通过两种控制系统的对比学习，得到以下结论：

(1) 继电器控制系统逻辑采用硬件接线，利用继电器机械触点的串联或并联等组合成控制逻辑，其连线多且复杂、体积大、功耗大。系统完成后，若再改变功能或增加功能较为困难。另外，继电器的触点数量有限，所以电器控制系统的灵活性和可扩展性受到很大限制。而 PLC 采用了计算机技术，其控制逻辑是以程序的方式存放在存储器中，要改变控制逻辑只需改变程序，因而很容易改变或增加系统功能。系统连线少、体积小、功耗小，PLC 的软继电器实质上是存储器单元的状态，所以软继电器的触点数量是无限的，PLC 系统的灵活性和可扩展性好。

(2) 继电器控制系统采用时间继电器的延时动作进行时间控制，时间继电器的延时时间定时精度不高。而 PLC 采用半导体集成电路作定时器，时钟脉冲由晶体振荡器产生，精度高，定时范围宽。用户可根据需要在程序中设定定时值，修改方便，且 PLC 具有计数功能，而继电器控制系统一般不具备计数功能。

(3) 由于电器控制系统使用了大量的机械触点，其存在机械磨损、电弧烧伤等现象。因为其寿命短，系统的连线多，所以可靠性和可维护性较差。而 PLC 控制系统中大量的开关动作由无触点的半导体电路完成，其寿命长，可靠性高。PLC 还具有自诊断功能，能查

出自身的故障，随时显示给操作人员，并能动态地监视控制程序的执行情况，为现场调试和维护提供了方便。

练习题

（1）电路分析接线题：某电动机正反转控制的继电器主控电路图如图3-43所示。

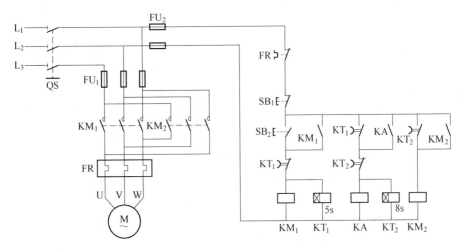

图3-43 电动机正反转控制的继电器主控电路图

1）分析该电路的控制过程，并将相应的元件功能填入表3-11中。

表3-11 电路元件功能表

序号	元件	功能	序号	元件	功能
1	SB_1		4	KM_2	
2	SB_2		5	KA_1	
3	KM_1		6	KA_2	

2）分别点动 SB_2 和 SB_1 时，电动机将如何工作，为什么？

3）完成该电路的接线，操作相应的按钮，对照自己的分析是否正确。

（2）电动机正反 PLC 主控电路接线与编程：在完成电动机正反转 PLC 控制系统的主电路设计和 PLC 外部电路接线图设计基础上，请按照以下设计要求的编写程序并下载到 PLC 中进行调试：

1）点动 SB_1，电动机延时 2s 正转，运行 3s 后停止；点动 SB_2，电动机立即反转启动，运行 5s 后停止。SB_3 为急停按钮。

2）点动 SB_1，电动机正转 10s 停 5s，然后反转。点动 SB_2，电动机延时 2s 停止，SB_3 为急停按钮。

3）点动 SB_1，电动机正转 5s 停 2s，再反转 6s 停 3s，依次循环。若点动 SB_2，则电动机延时 3s 停止，SB_3 为急停按钮。

4）SB_1 为模式选择按钮，SB_2 为启动按钮，SB_3 为急停按钮。启动前必须先进行模式选择：点动一次 SB_1，电动机运行模式为正转立即启动选择；点动两次 SB_1，电动机运行模式为延时 3s 正转启动选择；点动三次 SB_1，电动机运行模式为反转立即启动选择；点动四次 SB_1，电动机运行模式为反转运行 5s 后停止。运行模式选择完毕，点动 SB_2 启动电动机。SB_3 为急停按钮。SB_1 的点动次数在点动 SB_2 后清零。

（3）生产车间换气系统模拟控制 PLC 编程与调试：某企业生产车间的换气系统由 1 号、2 号两台风机以及一个状态指示灯组成。其中，1 号为主风机，2 号为备用风机，每台风机均有各自的启动按钮和停止按钮。整个换气系统有启动按钮和急停按钮。控制要求如下：

1）调试运行：每台风机均能单独启动和停止。

2）正常运行：点动系统启动按钮，1 号风机立即启动运行，运行指示灯由熄灭到常亮。在运行模式下点动停止按钮，风机系统立即停止，运行指示灯熄灭。

3）维护运行：在正常运行状态下，点动 1 号风机停止按钮，使其由工作状态切换到停止运行状态进行维护，运行指示灯熄灭，2 号风机延时 5s 自动启动运行。2 号风机运行后，指示灯以 1Hz 的频率闪烁；1 号风机完成维护后，点动 2 号风机的停止按钮，2 号风机立即停止，运行指示灯熄灭，1 号风机延时 3s 自动启动运行投入正常工作，指示灯变为常亮。

根据控制要求完成如下任务：

1）根据控制要求完成 PLC 的 I/O 地址分配。

2）设计 PLC 的外部硬件接线图。

3）编写 PLC 程序并运行调试。

项目 4 三相异步电动机的变频器控制

变频器是一种采用变频驱动技术改变交流电动机工作电压的频率和幅度来平滑控制交流电动机速度及转矩的可调速驱动装置,可以满足交流电动机无级调速的广泛需求。由于变频器在调速、节能方面的优异表现,其广泛应用于机床、自动化生产线、楼宇电梯等各种机械设备控制领域。本项目主要学习如何使用德国西门子公司的 SINAMICS G120C PN 变频器来控制三相异步电动机的运行。根据控制的手段不同,该项目包括利用 G120C 变频器操作面板控制三相异步电动机的运行和使用 G120C PN 变频器、S7-1200 PLC 控制三相异步电动机的运行两个任务,能够由浅入深逐步掌握变频器的使用。

任务 4.1 利用变频器面板控制三相异步电动机的运行

4.1.1 任务描述

SINAMICS G120C 变频器配备的 BOP-2 基本操作面板可以用于对变频器的调试、运行监控及输入参数的设置。通过液晶屏显示的菜单导航和相关参数进行变频器的调试,通过导航键可以方便地对变频器进行本地控制,一个专门的按键就可以完成手动/自动功能的切换。

本任务利用 SINAMICS G120C 变频器的 BOP-2 基本操作面板,完成变频器常用参数的设置,以及完成对三相异步电动机的启停、正反转运行和调速的控制。

4.1.2 任务目标

(1) 理解变频器的基本结构组成与工作原理。
(2) 熟悉西门子 G120C 变频器的操作面板和接线端子的功能。
(3) 掌握西门子 G120C 变频器的简单安装与接线方法。
(4) 掌握西门子 G120C 变频器常用参数的设置方法。
(5) 使用西门子 G120C 变频器的 BOP-2 基本操作面板控制三相异步电动机的启动、停止、正反转运行和调速控制。

4.1.3 任务相关知识

4.1.3.1 变频器概述

变频器的定义及作用

变频器(Inverter)是应用变频技术与微电子技术,通过改变电动机工作电源频率方式来控制交流电动机的电力控制设备。几款常见变频器的外观如图 4-1 所示。

图 4-1 几款常见变频器的外观

变频器的主要作用是通过改变电动机的供电频率，从而调节负载，降低功耗，减少能源损耗，延长设备的使用寿命，同时还能提高生产设备的自动化程度。目前，变频器已广泛应用于轻、重工业的生产以及人们的日常生活中。

变频器的结构组成及分类

● 变频器的结构组成

变频器由主电路和控制电路组成，主电路包括整流电路、滤波电路和逆变电路。变频器的结构框图如图 4-2 所示。

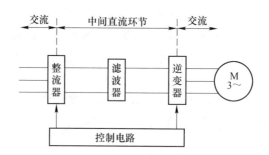

图 4-2 变频器的结构框图

（1）变频器的主电路：整流电路由三相全波整流桥组成，它的作用是对工频的外部电源进行整流，产生脉动的直流电压，给逆变电路和控制电路提供所需要的直流电源。滤波电路是对整流电路的输出进行平滑，以保证逆变电路和控制电路能够得到质量较高的直流电源。逆变电路是将直流中间电路输出的直流电源转换为频率和电压都可调的交流电源。

（2）变频器的控制电路：变频器的控制电路由运算电路、检测电路、控制信号的输入与输出电路和驱动电路等构成，其主要功能是完成对逆变电路的开关控制，对整流电路的电压控制以及完成各种保护功能。

● 变频器的分类

变频器可以分为：

（1）按输入电压等级分类。变频器按输入电压等级可分为低压变频器和高压变频器。

低压变频器常见的有单相220V变频器、三相220V变频器和三相380V变频器；高压变频器常见有6kV、10kV变频器。

（2）按变换频率的方法分类。变频器按频率变换的方法分为交—交式变频器和交—直—交式变频器。

交—交式变频器可将工频交流电直接转换成频率和电压均可控制的交流电（故称直接式变频器）。这种变频器输出的最高频率一般只能达到电源频率的1/3~1/2，适用于低频大容量的调速系统。

交—直—交式变频器可先把工频交流电通过整流装置转变成直流电，然后再把直流电变换成频率和电压均可调节的交流电（故称间接式变频器）。目前广泛应用的通用型变频器都是交—直—交式变频器。

（3）按变频器用途分类。按变频器用途可分为通用变频器和专用变频器。

通用变频器能够适用于所有负载的变频器。专用变频器是在通用变频器的基础上，针对变频器所拖动负载的特性进行了一系列专门的优化，具有参数设置更简单、节能调速效果更佳等一系列的优点。常见的专用变频器有风机专用变频器、水泵专用变频器、电梯专用变频器等。

4.1.3.2 SINAMICS G120C 变频器简介

SINAMICS G120C 变频器是德国西门子公司生产的一款用于控制三相交流电动机转速的一体式紧凑型变频器，是SINAMICS驱动家族的成员之一。其可满足传送带、搅拌机、挤压机、水泵、风机、压缩机以及一些基本的物料处理机械等众多应用的需求，功率范围0.55~132kW，支持PROFINET、EtherNet/IP、PROFIBUS、USS/Modbus RTU 等多种通信方式。该产品能够内置于控制箱和开关柜中，从而节省空间。其外观如图4-3所示。

图4-3 SINAMICS G120C 变频器的外观

SINAMICS G120C 变频器由控制单元（CU）和功率模块（PM）组成。下面分别介绍这两部分的接线端子的功能及接线方式。

SINAMICS G120C 变频器控制单元端子介绍

SINAMICS G120C 变频器控制单元的端子功能与接线方式如图4-4所示。

● 电源端子

端子1（+10V OUT）、2（GND）是变频器为用户提供的一个高精度的10V直流电源。

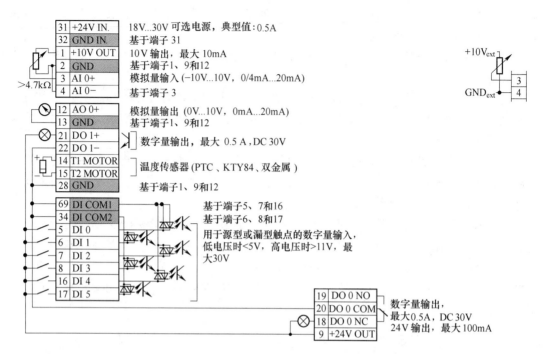

图 4-4 SINAMICS G120C 变频器的端子与接线方式

端子 9（+24V OUT）、28（GND）是变频器的内部 24V 直流电源，最大输出电流为 10mA，可供数字量输入端子使用。

端子 31（+24V IN）、32（GND IN）是外部接入的 24V 直流电源，用户为变频器的控制端提供 24V 直流电源。

- 数字量端子

端子 5（DI 0）、6（DI 1）、7（DI 2）、8（DI 3）、16（DI 4）、17（DI 5）为用户提供了 6 个完全可编程的数字输入端，数字输入信号经光电隔离输入 CPU，对电动机进行正反转、正反点动、固定频率设定值控制等。

端子 18（DO 0 NC）、19（DO 0 NO）、20（DO 0 COM）及 21（DO 1 POS）、22（DO 1 NEG）数字量输出端，其中 18、19、20 为继电器型输出；21、22 为晶体管型输出。

- 模拟量端子

端子 3（AI 0+）、4（AI 0-）为用户提供了模拟电压给定输入端，可作为频率给定信息，经变频器内模/数转换器，把模拟量转换成数字量，传输给 CPU。

端子 12（AO 0+）、13（GND）为模拟量输出端，可为仪器仪表或控制器输入端提供标准的直流模拟信号。

- 通信端子

控制单元为用户提供了 2 个 ProfiNet 通信接口，以便和其他控制器进行数据通信。

- 保护端子

端子 14（T1 MOTOR）、15（T2 MOTOR）为电动机过热保护输入端，当电动机过热时给 CPU 提供一个触发信号。

● 公共端子

端子 34（DI COM2）、69（DI COM1）为数字量公共端子，在使用数字量输入时，必须将对应的公共端子与 24V 电源的负极端相连。

练一练

观察 G120C 变频器的外观，查看在变频器右侧面的产品标签上的参数。卸下 BOP-2 基本操作面板，打开 BOP-2 下方的盖子，观察变频器的接线端子的位置和标号，如图 4-4 所示，思考如何接线。

SINAMICS G120C 变频器功率单元的连接

G120C 变频器与输入电源、电动机的连接如图 4-5 所示。图中的输入、输出电抗器和制动电阻都是根据变频器需要可以选配的部件。

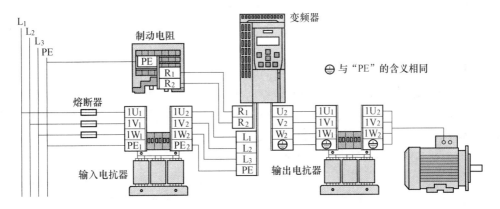

图 4-5　SINAMICS G120C 变频器与电源、电动机的连接

输入、输出电抗器和制动电阻的作用及接线如下：

（1）制动电阻：制动电阻可为变频器实现对高转动惯量负载的有效制动。连接到位于 G120C 变频器底部的 R_1、R_2 端，如图 4-6 所示。

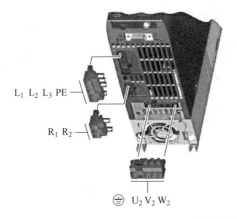

图 4-6　SINAMICS G120C 变频器与电源、电动机的接口端子

（2）输入电抗器：输入电抗器可增强变频器对过压、谐波和换向扰动的防护。进线端

通过熔断器连接到三相电源的 L_1、L_2、L_3，如图 4-5 所示；出线端连接到位于 G120C 变频器底部的 L_1、L_2、L_3 和 PE 端，如图 4-6 所示。

（3）输出电抗器：输出电抗器用于延长变频器的有效传输距离，有效抑制变频器的 IGBT 模块开关时产生的瞬间高压。进线端连接到位于 G120C 变频器底部 U_2、V_2、W_2 及 PE 端子，如图 4-6 所示；出线端连接到三相交流电动机的接线端子，如图 4-5 所示。

在变频器安装前，制动电阻、输入电抗器、输出电抗器及电源电缆、信号电缆的选型与变频器的安装方法请参考产品安装手册。在安装过程中，变频器必须保证可靠接地，在进行电缆连接或改动接线时必须断开电源，电源线与信号线必须分开敷设，以防止电磁干扰影响设备的正常工作。

练一练

将 G120C 变频器的 L_1、L_2、L_3 三相电源线及 PE 地线连接到输入电源端，将三相异步电动机连接到变频器 U_2、V_2、W_2 及 PE 端子上。注意三相异步电动机绕组的接线方式。

4.1.3.3 SINAMICS G120C 变频器的调试

第一次使用新的变频器或者更换变频器所驱动的电动机时，需要把电动机的铭牌参数数据和一些基本驱动控制参数输入到变频器中，以便更好地驱动电动机运行。下面利用 BOP-2 基本操作面板来对变频器进行设置与调试工作。

BOP-2 基本操作面板的认识与安装

BOP-2 基本操作面板如图 4-7 所示，其可以用于对变频器的调试、运行监控及某个参数的设置，并可通过菜单导航和两行参数来显示变频器的调试过程。

将 BOP-2 基本操作面板插到变频器上，操作步骤如图 4-8 所示。

图 4-7　BOP-2 基本操作面板　　　　图 4-8　安装 BOP-2 基本操作面板

1—菜单条指出的当前选中的菜单功能；
2—提供的有关选定功能的信息或显示实际值；3—显示数值

按如下步骤将 BOP-2 基本操作面板插到变频器上：

（1）拆下变频器的保护盖。

（2）将 BOP-2 下边缘插入变频器对应的凹槽中。

（3）将 BOP-2 推入变频器，直到听到 BOP-2 在变频器外壳上卡紧的声音，成功插入了 BOP-2。

BOP-2 上主要有 7 个功能按键，其功能见表 4-1。

表 4-1 BOP-2 基本操作面板按键的功能

按键	功能
OK	〈OK〉键具有以下功能： (1) 浏览菜单时，按〈OK〉键确定选择一个菜单项。 (2) 进行参数操作时，按〈OK〉键允许修改参数。再次按〈OK〉键，确认输入的值并返回上一页。 (3) 在故障屏幕，该键用于清除故障
▲	向上〈UP〉键具有以下功能： (1) 当浏览菜单时，该键将光标移至向上选择当前菜单下的显示列表。 (2) 当编辑参数值时，按下该键增大数值。 (3) 如果激活 HAND 模式和点动功能，同时长按向上键和向下键有以下作用： 　1) 当反向功能开启时，关闭反向功能； 　2) 当反向功能关闭时，开启反向功能
▼	向下〈DOWN〉键具有以下功能： (1) 当浏览菜单时，该键将光标移至向上选择当前菜单下的显示列表。 (2) 当编辑参数值时，按下该键减小数值。编辑参数值时减小显示值
ESC	〈ESC〉键具有以下功能： (1) 如果按下时间不超过 2s，则 BOP-2 返回到上一页。如果正在编辑数值，新数值不会被保存。 (2) 如果按下时间超过 3s，则 BOP-2 返回到状态屏幕。在参数编辑模式下使用〈ESC〉键时，除非先按确认键，否则数据不能被保存
I	开机〈ON〉键具有以下功能： (1) 在 AUTO 模式下，开机键未被激活，即使按下它也会被忽略。 (2) 在 HAND 模式下，变频器启动电机；操作面板屏幕显示驱动运行图标
O	关机〈OFF〉键具有以下功能： (1) 在 AUTO 模式下，关机键不起作用，即使按下它也会被忽略。 (2) 如果按下时间超过 2s，变频器将执行 OFF2 命令；电机将关闭停机。 (3) 如果按下时间不超过 3s，变频器将执行以下操作： 　1) 如果两次按关机键不超过 2s，将执行 OFF2 命令。 　2) 如果在 HAND 模式下，变频器将执行 OFF1 命令；电机将在参数 P1121 中设置的减速时间内停机
HAND/AUTO	手动/自动〈HAND/AUTO〉键切换 BOP-2（HAND）和现场总线（AUTO）之间的命令源，其具有以下功能： (1) 在 HAND 模式下，按〈HAND/AUTO〉键将变频器切换到 AUTO 模式，并禁用开机和关机键。 (2) 在 AUTO 模式下，按〈HAND/AUTO〉键将变频器切换到 HAND 模式，并启用开机和关机键。 (3) 在电动机运行时也可切换 HAND 模式和 AUTO 模式

练一练

(1) 练习拆卸、安装 BOP-2 基本操作面板。

(2) 接通变频器电源,熟悉面板按键的功能。

BOP-2 基本操作面板的菜单结构与功能

BOP-2 基本操作面板菜单结构及功能如图 4-9 所示。

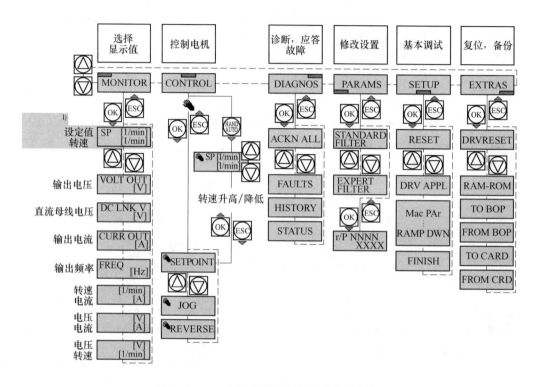

图 4-9 BOP-2 基本操作面板的菜单及功能

菜单功能简单描述见表 4-2。

表 4-2 BOP-2 基本操作面板菜单的功能描述

菜单	功能描述
MONITOR	监视菜单:显示电动机的实际转速、实际输出电压和电流值
CONTROL	控制菜单:允许用户访问变频器的设定值、点动和反向功能
DIAGNOS	诊断菜单:故障报警和控制字、状态字的显示
PARAMS	参数菜单:允许用户查看和更改变频器参数值
SETUP	设置菜单:允许用户执行变频器的标准调试
EXTRAS	附加菜单:允许用户执行恢复出厂默认设置和数据备份

BOP-2 基本操作面板会在显示屏左侧显示各种图标,显示变频器的实际状态。图标的描述见表 4-3。

表 4-3　BOP-2 基本操作面板图标描述

图标	功能	状态	备　注
✋	命令源	手动模式	当 HAND 模式启用时，显示该图标；当 AUTO 模式启用时，无图标显示
⊕	变频器状态	变频器和电机运行状态	该图标是静止的，不旋转
JOG	点动功能	点动功能激活	激活变频器点动功能
⊗	故障/报警	故障或报警等待闪烁的符号=故障 稳定的符号=报警	故障状态下，会闪烁，变频器会自动停止。静止图标表示处于报警状态，不会停止变频器运行

使用 BOP-2 基本操作面板修改参数

修改参数值是在菜单 PARAMS 中进行，具体操作步骤如下：

（1）选择参数号。当显示的参数号闪烁时，按〈UP〉键或〈DOWN〉键选择所需的参数号；按下〈OK〉键进入参数，显示当前参数值。

（2）修改参数值。当显示的参数值闪烁时，按〈UP〉键或〈DOWN〉键调整参数值；按下〈OK〉键保存参数值。

练一练

给变频器通电，熟悉面板的菜单操作方法，练习修改参数值。

使用 BOP-2 基本操作面板对变频器快速调试

变频器的快速调试是通过设置电动机基本参数，达到简单快速运转电动机的一种操作模式。

进行快速调试的前提条件如下：

（1）接通电源。

（2）操作面板显示电动机转速的设定值和实际值。

按如下步骤可以进行快速调试：

（1）按下〈ESC〉键。

（2）按下〈UP〉键，直到 BOP-2 显示"SETUP"菜单。

（3）在"SETUP"菜单中按下〈OK〉键，以起动快速调试。

（4）如果希望在快速调试前恢复所有参数的出厂设置，过程如下：

1）按下〈OK〉键；

2）使用箭头键切换：NO → YES；

3）按下〈OK〉键。

屏幕显示 BUSY，一段时间后变为 DONE，最后显示 DRV_APPL　P96_。

（5）选择了应用等级后，变频器会为电动机匹配合适的缺省设置：

1）0　EXPERT；

2）1　STANDAR（Standard Drive Control，标准驱动控制）；

3）2　DYNAMIC（Dynamic Drive Control，动态驱动控制）。

这里选择 STANDAR。变频器快速调试过程中的相关参数设置见表 4-4。

表 4-4 变频器的快速调试中参数设置

屏幕显示的参数	功　能	设　置
EUR/USA P100	设置电动机标准，系统提供 3 个选项：0 KW 50Hz，1 HP 60Hz，2 KW 60Hz	KW 50Hz IEC
INV VOLT P210	设置变频器的输入电压	380
MOT TYPE P300	设置电动机类型	INDUCT（表示属于第三方异步电动机）
MOT CODE P301	电动机代码：如果使用电动机代码选择电动机类型，需要输入；如果不知道电动机代码，必须将电动机代码设为 0	0
87 HZ	电动机 87Hz 运行：只有选择了 IEC 作为电动机标准，BOP-2 才显示该步骤，可选 NO 或 YES。87Hz 可以使电动机的运行速度达到正常速度的 1.73 倍	No
MOT VOLT P304	电动机额定电压	380
MOT CURR P305	电动机额定电流	根据电动机铭牌设置
MOT POW P307	电动机额定功率	根据电动机铭牌设置
MOT FREQ P310	电动机额定频率	50
MOT RPM P311	电动机额定转速	根据电动机铭牌设置
MOT COOL P335	电动机冷却方式，系统提供 4 个选项： （1）0 SELE 自然冷却； （2）1 FORCED 强制冷却； （3）2 LIAUID 液冷； （4）128 NO FAN 无风扇	128 NO FAN
TEC APPL P501	选择应用，系统提供两个选项： （1）0 VEC STD：恒定负载，典型应用为输送驱动； （2）1 PUMP FAN：取决于转速的负载，典型应用为泵和风机	0
MAc PAr P15	选择与应用相适宜的变频器接口的缺省设置	1
MIN RPM P1080	电动机的最小转速，默认值为 0	根据电动机铭牌设置
MAX RPM P1082	电动机的最大转速，默认值为 1500r/min	根据电动机铭牌设置

续表 4-4

屏幕显示的参数	功　能	设　置
RAMP UP P1120	电动机升速时间，默认值为 10s	10
RAMP DWN P1121	电动机减速时间，默认值为 10s	10
OFF3 RP P1135	符合 OFF3 指令的降速时间，默认值为 0	0
MOT ID P1900	电动机数据检测，选择变频器测量所连电动机数据的方式，系统提供 6 个选项：0 OFF，1 STIL ROT，2 STIL，3 ROT，11 ST RT OP，12 STILL OP	0（电动机数据未测量）
FINISH	结束快速调试	使用箭头键切换 NO→YES，按下〈OK〉键

成功输入变频器快速调试需要的所有数据，此时面板显示 BUSY，变频器进行参数计算，计算完成变频器显示 DONE 界面，随后光标返回到 MONITOR 菜单。

如果在快速调试中设置 P1900 不为 0，在快速调试后屏幕会显示报警 A07991，红色故障 LED 灯闪烁，提示激活电动机数据检测，等待启动命令。

在快速调试完成后，如果还有其他参数需要修改，或快速调试中电动机的部分参数还需要修改，可直接在专家参数列表中修改。打开专家参数列表，首先设置快速调试参数 P0010=1，只有在 P0010=1 的情况下才能修改电动机的相关参数。参数修改完成后，必须使 P0010=0。

练一练

按照变频器所连接三相异步电动机的铭牌数据，在手动模式下使用 BOP-2 基本操作面板对变频器进行快速调试的相关设置。

使用 BOP-2 更改设置

变频器设置是通过更改变频器中的参数值来修改的，参数分为可写参数和只读参数两种。变频器只允许更改可写参数，可写参数以"P"开头，如 P45；只读参数以"r"开头，只读参数的值不允许更改，如 r2。

根据以下步骤使用 BOP-2 更改可写参数，以修改参数号 P45 的内容为例，如图 4-10 所示。

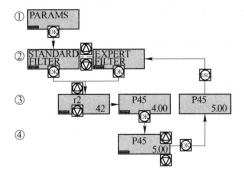

图 4-10　使用 BOP-2 更改设置操作步骤

从图 4-10 可以看出，其步骤分别为：

（1）选择 PARAMS 菜单，按下〈OK〉键。

（2）使用箭头键选择参数筛选条件 STANDARD FILTER 或 EXPERT FILTER，按下〈OK〉键。其中，STANDARD：变频器只显示重要参数；EXPERT：变频器显示所有参数。

（3）屏幕上显示最后一次设置的参数内容。使用箭头键选择需要的可写参数号，如 P45。此时数字"45"处于闪烁状态，按下〈OK〉键后，P45 的内容"4.00"开始闪烁，说明其处于可修改的状态。

（4）使用箭头键设置可写参数值，如"4.00"修改为"5.00"，确定数值后，按下〈OK〉键接受该值。

使用 BOP-2 成功更改了可写参数，屏幕出现 BUSY。

在带下标的参数上，一个参数号有多个参数值，每个参数值有一个单独的下标。

以修改 P840 的下标参数为例，操作步骤如图 4-11 所示。

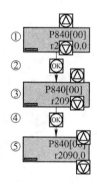

图 4-11　更改带下标参数的操作步骤

从图 4-11 可以看出，其操作步骤分别为：

（1）选择参数号 P840，"840"闪烁。

（2）按下〈OK〉键，[00] 开始闪烁，说明可以修改此下标。

（3）使用箭头键设置参数下标，可以把 [00] 修改为 [01]。

（4）按下〈OK〉键。

（5）成功更改了带下标的参数。

4.1.3.4　SINAMICS G120C 变频器的预定义接口宏

G120C 变频器的预定义接口宏概述

SINAMICS G120C 变频器集成了多种预定义接口宏功能，使用者可以直接调用。每种宏对应着一种接线方式，选择其中一种宏后变频器会自动设置与其接线方式相对应的一些参数，这样极大方便了用户的快速调试，从而提高了调试效率。

通过参数 P0015 修改宏编号。只有在 P0010 = 1 时，才能修改 P0015 的参数值。修改 P0015 的步骤如下：

（1）设置 P0010 = 1；

（2）修改 P0015 的参数值；

(3) 设置 P0010=0。

G120C PN 变频器的 17 种宏功能见表 4-5。

表 4-5 G120C PN 变频器的 17 种宏功能

宏编号	宏功能描述
1	双线制控制，有两个固定转速
2	单方向两个固定转速，带安全功能
3	单方向四个固定转速
4	现场总线
5	现场总线，带安全功能
6	现场总线，带两项安全功能
7	现场总线和点动之间切换
8	电动电位器（MOP），带安全功能
9	电动电位器（MOP）
12	双线控制 1
13	端子启动模拟量给定，带安全功能
14	现场总线和电动电位器（MOP）切换
15	模拟量给定和电动电位器（MOP）切换
17	双线控制 2
18	双线控制 3
19	三线控制 1
20	三线控制 1

在下面的变频器控制电动机的多段速运行，就是使用宏编号为 1 的预定义接口宏。变频器外接电位器调节电动机转速，就是使用宏编号为 13 的预定义接口宏。

G120C 变频器的预定义接口宏举例

● 预定义接口宏 1

设置 G120C 变频器参数 P0015=1，端子功能如图 4-12 所示。

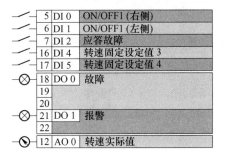

图 4-12 预定义接口宏 1 的端子功能

该宏的功能为双线制控制，有两个固定转速。DI 0 定义为正转的启/停功能；DI 1 定义为反转的启/停功能；DI 4 定义为转速固定设定值 3，该转速设定值在 P1003 中设置；

DI 5 定义为转速固定设定值 4,该转速设定值在 P1004 中设置;DI 4 和 DI 5 都接通时,变频器以"转速固定设定值 3+转速固定设定值 4"运行。

- 预定义接口宏 13

设置 G120C 变频器参数 P0015=13,端子功能如图 4-13 所示。

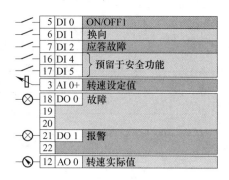

图 4-13 预定义接口宏 13 的端子功能

该宏的功能为端子起动模拟量给定,带安全功能。DI 0 定义为启/停功能;DI 1 定义为换向功能;AI 0+定义为外接模拟量来调整转速设定值。

4.1.3.5 SINAMICS G120C 变频器的指令源和设定值源

通过预定义接口宏可以定义变频器用什么信号控制启动,由什么信号来控制输出频率。

命令源

命令源是指变频器收到控制命令的接口。在设置预定义接口宏 P0015 时,变频器会自动对命令源进行定义。表 4-6 所列举的参数设置中 r722.0、r722.2、r722.3、r2090.0、r2090.1 均为命令源。

表 4-6 命令源示例

参数号	参数值	说　　明
P0840	722.0	将数字输入 DI 0 定义为启动命令
	2090.0	将现场总线控制字 1 的位 0 定义为启动命令
P0844	722.2	将数字输入 DI 2 定义为 OFF2
	2090.1	将现场总线控制字 1 的位 1 定义为 OFF2 命令
P2013	722.3	将数字输入 DI 3 定义为故障复位

设定值源

设定值源是指变频器收到设定值的接口。在设置预定义接口宏 P0015 时,变频器会自动对设定值源进行定义。主设定值 P1070 的常用设定值源见表 4-7。其中,r1050、r755.0、r1024、r2050.1、r755.1 均为设定值源。

表 4-7 设定值源示例

参数号	参数值	说明
P1070	1050	将电动电位器作为主设定值
	755.0	将模拟量输入 AI 0 作为主设定值
	1024	将固定转速作为主设定值
	2050.1	将现场总线过程数据作为主设定值
	755.1	将模拟量输入 AI 1 作为主设定值

4.1.4 任务实施

在 XK-SX2C 高级维修电工实训台上，使用跨接线完成变频器与三相电源之间的连接，完成变频器与三相异步电动机的连接。在确保连接正确的基础上，变频器通电。根据电动机的铭牌数据，使用 BOP-2 面板完成变频器的快速调试，并在此基础上，完成以下操作任务。

4.1.4.1 操作变频器 BOP-2 面板控制电动机的启停运行

任务要求

操作 BOP-2 面板实现对三相异步电动机的启动、停止及连续运行，并能调节三相异步电动机的转速。完成后对本项目进行评分，详见表 4-10。

操作步骤

其操作步骤分别为：

（1）在 BOP-2 面板上，按下〈HAND/AUTO〉按键，BOP-2 屏幕上显示 HAND 图标，变频器进入手动运行模式。

（2）按下〈ESC〉键，进入菜单选择。

（3）按下〈UP〉键或〈DOWN〉键，将菜单条移至 CONTROL，然后按〈OK〉键。

（4）屏幕上显示 SET POINT，再按下〈OK〉键，显示 SP，此时显示的电动机转速值为 0。

（5）按下 UP 键，调整电动机的转速设定值，如 1000r/min。

（6）按下〈ON〉键启动电动机，电动机转速逐渐上升到设定值。

（7）此时，按向上或向下箭头键，改变电动机转速设定值，从而改变电动机的转速。

（8）此时，按下〈OK〉键，可以切换显示电动机运行的电压、电流、频率等数值。

（9）按下〈OFF〉键，电动机停止运行，转速逐渐下降到 0。

4.1.4.2 操作变频器 BOP-2 面板控制电动机的点动运行

任务要求

操作 BOP-2 面板实现对三相异步电动机点动运行。完成后对本项目进行评分，详见表 4-10。

操作步骤

其操作步骤分别为：

(1) 在 BOP-2 面板上，按下〈HAND/AUTO〉按键，BOP-2 屏幕上显示 HAND 图标，变频器进入手动运行模式。

(2) 按下〈ESC〉键，进入菜单选择。

(3) 按下〈UP〉键或〈DOWN〉键，将菜单条移至 CONTROL，然后按〈OK〉键。

(4) 屏幕上显示 SET POINT，再按下〈OK〉键，显示 SP，此时显示的电动机转速值为 0。

(5) 按下〈UP〉键，设定电动机的转速数值，如 1000r/min。

(6) 按下〈ESC〉键返回到屏幕显示 SET POINT。

(7) 按下〈DOWN〉键，屏幕显示 JOG，然后按〈OK〉键，屏幕显示 OFF；按下〈DOWN〉键，显示 ON，再按〈OK〉键，设置成功，屏幕左下角显示 JOG，电动机运行选择为点动模式。

(8) 一直按住 ON 键启动电动机，电动机转速逐渐上升，到某个转速值停止，没有达到 1000r/min，因为点动操作的电动机转速不是使用 SET POINT 设置的，点动转速的数值在参数号 P1058 中设置，出厂时默认为 150r/min。

(9) 松开〈ON〉键，电动机停止运行，转速逐渐下降到 0。

4.1.4.3 操作变频器 BOP-2 面板控制电动机的反向运行

任务要求

操作 BOP-2 面板实现三相异步电动机的反向运行。完成后对本项目进行评分，详见表 4-10。

操作步骤

其操作步骤分别为：

(1) 在 BOP-2 面板上，按下〈HAND/AUTO〉按键，BOP-2 屏幕上显示 HAND 图标，变频器进入手动运行模式。

(2) 按下〈ESC〉键，进入菜单选择。

(3) 按下〈UP〉键或〈DOWN〉键，将菜单条移至 CONTROL，然后按〈OK〉键。

(4) 屏幕上显示 SET POINT，再按下 OK 键，显示 SP，此时显示的电动机转速值为 0。

(5) 按下〈UP〉键，设定电动机的转速数值，如 1000r/min。

(6) 按下〈ON〉键启动电动机，电动机转速逐渐上升到设定值，观察并记住电动机运行的方向。

(7) 按下〈ESC〉键，返回到屏幕显示 SET POINT。

(8) 按下〈DOWN〉键两次，屏幕显示 REVERSE，然后按〈OK〉键，屏幕显示 OFF；按下〈DOWN〉键，显示 ON，再按〈OK〉键，设置成功，电动机运行选择为反向模式。

(9) 电动机在原有运行方向上逐渐减速直至为 0。此时，电动机开始反向，直至加速到设定转速数值。

(10) 按下〈OFF〉键，电动机停止运行，转速逐渐减低到 0。

另外一种改变三相异步电动机运行方向的操作步骤分别为：

(1) 在电动机处于运行的状态下，屏幕上显示 SET POINT 菜单，再按下〈OK〉键，显示 SP 的数值。

（2）如果 SP 显示的转速数值为正值，按住〈DOWN〉键减小转速设定值。当设定值出现负号的时候，电动机的转向变为反向。

（3）同理，当 SP 显示的转速数值为负值时，按住〈UP〉键。当设定值的负号消失变为正值时，电动机变为正向运行。

以上 3 个操作任务是利用 BOP-2 完成变频器的简单设置和操作控制。下面的操作任务将利用数字量端子连接多个开关，实现对三相异步电动机的多段速运行。

4.1.4.4 变频器控制三相异步电动机的多段速运行

任务要求

多段速运行又称固定频率运行。通过在 G120C 变频器的数字量端子外接开关，实现电动机分别以 400r/min、800r/min 和 1200r/min 3 个转速运行，并可以通过开关在 3 个转速之间切换。

完成后对本项目进行评分，详见表 4-10。

任务分析

在设置 P1000=3 的条件下，通过 G120C 变频器的数字量端子外接开关，选择固定设定值的组合，实现电动机多段速运行。其中，有两种固定设定值模式，即直接选择和二进制选择。G120C 变频器的数字量输入端子 5、6、7、8、16、17 为用户提供了 6 个完全可编程的数字量输入端子，其输入信号可以来自外部的按钮、继电器或晶体管的输出信号。端子 9、28 是变频器内置的 24V 的直流电源，提供数字量的输入所需的直流电源。通过在数字量输入端外接按钮，并对相关参数进行设置，实现对电动机以多个段速运行及启/停控制。

操作步骤

其操作步骤分别为：

（1）变频器接线：按照图 4-14 完成变频器和数字量端子的接线工作。

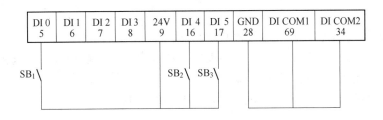

图 4-14　G120C 变频器数字量端子的接线

如图 4-14 所示，在数字量输入端 5、16、17 外接 3 个按钮，这些按钮可通过参数设置，定义为正转启/停、固定频率运行控制等功能。

（2）变频器参数设置：实现电动机以 400r/min、800r/min 和 1200r/min 3 个转速运行的变频器，参数设置见表 4-8。

表 4-8 变频器参数设置表

参数号	参数值	功 能	备注
P0015	1	预定义宏参数。宏编号为 1，宏功能为双线制控制，有两个固定频率 注意：修改 P0015 时，必须在 P0010=1 时更改，修改后使 P0010=0	默认值
P0840	722.0	将 DI 0 作为启动信号，r722.0 为 DI 0 状态的参数，SB$_1$ 被定义为电动机的正转启/停控制功能	
P1000	3	P1000 为频率控制源参数，P1000=3 为固定频率运行	
P1016	1	P1016=1 定义为固定频率运行采用直接选择模式。此模式下，P1003 为固定转速 3，P1004 为固定转速 4，DI 4、DI 5 都接通时，变频器以"固定转速 3+固定转速 4"的转速运行	
P1022	722.4	将 DI 4 作为固定转速 3 的选择信号，r722.4 为 DI 4 状态的参数，SB$_2$ 作为一个选择信号的输入	
P1023	722.5	将 DI 5 作为固定转速 4 的选择信号，r722.5 为 DI 5 状态的参数，SB$_3$ 作为一个选择信号的输入	
P1003	400	定义固定转速 3，单位为 r/min，这里设置为 400r/min	需设置
P1004	800	定义固定转速 4，单位为 r/min，这里设置为 800r/min	
P1070	1024	定义固定转速值作为主设定值	默认值

根据表 4-8 的相关参数设置，SB$_1$ 被定义为电动机正转启停控制按钮；SB$_2$ 被定义为以 400r/min 转速运行的控制按钮；SB$_3$ 被定义为以 800r/min 转速运行的控制按钮。同时按下 SB$_2$ 和 SB$_3$ 按钮，电动机以 1200r/min 转速运行。

任务演示

其演示步骤分别为：

（1）按住 SB$_1$，电动机接通电源。

（2）按下 SB$_2$，电动机以 400r/min 转速运行；释放 SB$_2$，电动机停止运行。

（3）按下 SB$_3$，电动机以 800r/min 转速运行；释放 SB$_3$，电动机停止运行。

（4）同时按下 SB$_2$ 和 SB$_3$，电动机以 1200r/min 转速运行；释放 SB$_2$、SB$_3$，电动机停止运行。

（5）释放 SB$_1$，电动机断开电源停止运行。

三相异步电动机的多段速运行是通过变频器的数字量输入端外接开关来实现的，但是电动机的转速只能在几种转速之间切换，而不能实现转速的连续变化。下面通过变频器的模拟量输入端外接电位器的方式实现电动机转速的连续调节。

4.1.4.5 G120C 变频器外接电位器调节三相异步电动机的转速

任务要求

在 G120C 变频器上使用外接的电位器实现三相异步电动机运行转速的实时连续调节。完成后对本项目进行评分，详见表 4-10。

任务分析

G120C 变频器通过外接电位器，产生的模拟电压信号送到模拟量输入端子 3（AI 0+）

作为频率给定信息，经变频器内模/数转换器把输入模拟量转换成数字量，传输给 CPU 来调节变频器输出电源的频率。G120C 变频器的端子 1、2 是变频器内置的 10V 的直流电源，作为电位器所需的直流电源，通过调节电位器的阻值来改变并获得连续变化的电压值，实现对电动机运行转速的连续调节。

操作步骤

其操作步骤分别为：

（1）变频器接线：按照图 4-15 完成变频器的模拟量端子、数字量端子的接线工作。

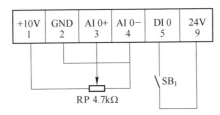

图 4-15　变频器端子接线图

在图 4-15 中，在数字量输入端 5 外接按钮 SB_1，通过参数设置，定义为正转起停功能。阻值为 4.7kΩ 的线绕电位器 RP，两端的直流电压取自 G120C 变频器内部的 10V 直流电源，电位器的可调端接入到变频器的端子 3，端子 4 与端子 2 连接 GND。另外，变频器控制单元正面保护盖后面的一个拨码开关拨到"U"的位置，表明是采用电压输入信号。

（2）变频器参数设置：实现电动机以转速连续调节运行的变频器参数设置见表 4-9。

表 4-9　变频器参数设置表

参数号	参数值	功　能
P0015	13	预定义宏参数。宏编号为 13，宏功能为端子启动由模拟量给定 注意：修改 P0015 时，必须在 P0010＝1 时更改，修改后使 P0010＝0
P0840	722.0	将 DI 0 作为启动信号，r722.0 为 DI 0 状态的参数，SB_1 被定义为电动机的正转启/停控制功能
P0756	0	模拟量输入类型选择： （1）0：0～10V；（2）1：2～10V；（3）2：0～20mA；（4）3：4～20mA；（5）4：-10～+10V； （6）8：没有连接传感器，这里选择 0
P0757	0	0V 对应频率为 0Hz，即对应电动机转速为 0r/min
P0758	0	
P0759	10	10V 对应频率为 50Hz，即对应电动机的转速为额定转速
P0760	100	
P1080	400	设置电动机的最小转速为 400r/min
P1082	1300	设置电动机的最大转速为 1300r/min

根据表 4-9 的相关参数设置，SB_1 被定义为电动机正转启停控制按钮，电动机启动运行后，转速在 400～1300r/min 连续调节。

任务演示

其演示步骤分别为：

（1）按住 SB_1，电动机接通电源，电动机启动运行。

（2）调节电位器旋钮，观察电动机的转速变化。利用 MINITOR 菜单观察变频器输出转速的变化范围，可以看到，转速在 400～1300r/min 连续调节。

（3）释放 SB_1，电动机逐渐停止运转。

4.1.5 任务评价

对变频器控制三相异步电动机的多段速运行操作任务的完成情况进行评分，并将评分结果填入表 4-10 中。

表 4-10 评分表

任务内容	考核要求	评分标准	配分	扣分	得分
电路设计	正确设计变频器控制线路接线图	（1）电气控制原理设计功能不全，每缺一项功能扣 5 分； （2）接线图表达不正确或画图不规范每处扣 2 分	10		
电路接线	变频器与三相电源；与三相异步电动机的接线；与数字量端子的接线	（1）变频器与三相电源的连接，与三相异步电动机的连接，每接错一根扣 10 分； （2）变频器数字量端子与按钮的连接，每接错一根扣 5 分	30		
变频器参数设置与调试	按照任务规定的控制要求进行变频器相应参数的设置，并运行调试达到任务要求	（1）变频器通电运行失败，每次扣 10 分； （2）变频器参数设置不全，每处扣 3 分； （3）参数设置错误每处扣 2 分； （4）没有设置参数扣 20 分； （5）变频器操作错误，每次扣 5 分	50		
安全文明生产	遵守 8S 管理制度，遵守安全管理制度	（1）未穿戴劳动保护用品，扣 10 分； （2）操作存在安全隐患，每次扣 5 分，扣完为止； （3）操作现场未及时整理整顿，每次发现扣 2 分，扣完为止	10		
总 分					

4.1.6 任务小结

任务实施过程记录单见表 4-11。

表 4-11 任务实施过程记录单

	任务实施过程记录单	
故障 1	故障现象	
	故障原因	
	排除过程	
故障 2	故障现象	
	故障原因	
	排除过程	
故障 3	故障现象	
	故障原因	
	排除过程	
	总　结	

4.1.7 任务拓展

任务要求：SINAMICS G120C 变频器的数字量输出端子可以用来指示三相异步电动机的运行状态，如：数字量输出 DO 0，其端子号为 18、19、20，属于继电器型输出，对应参数号为 P0730。在端子 19 和 20 之间接 24V 直流电源和指示灯，来显示电动机的运行状态。

请在变频器控制三相异步电动机的多段速运行操作任务的基础上，自行查询产品手册，重新设计电路，完成接线，完成相关参数的设置，实现电动机运行状态的指示。

任务 4.2　利用 PLC 和变频器控制三相异步电动机运行

4.2.1　任务描述

PLC 作为传统继电器的替代品，已经广泛应用于工业控制的各个领域。它可以通过软件来改变控制过程，且具有体积小、组装灵活、编程简单、抗干扰能力强及可靠性高的优点，非常适用于在恶劣工作环境下应用。当利用变频器构成自动控制系统进行控制时，许多情况是采用和 PLC 配合使用。

本任务利用西门子 S7-1215C PLC 和 SINAMICS G120C 变频器构成控制系统，完成对三相异步电动机的启停、正反转和多段速度运行的控制。

4.2.2　任务目标

（1）掌握 PLC 对西门子 G120 变频器控制的接线方法。

（2）掌握西门子 G120C 变频器的常用参数的设置。

（3）能够独立完成使用西门子 S7-1215C PLC 和 G120 变频器实现三相异步电动机的启停控制线路的设计、安装、编程调试与运行。

（4）能够独立完成使用西门子 S7-1215C PLC 和 G120 变频器实现三相异步电动机的

正反转控制线路的设计、安装、编程调试与运行。

(5) 能够独立完成使用西门子 S7-1215C PLC 和 G120 变频器实现三相异步电动机的多段速度控制线路的设计、安装、编程调试与运行。

4.2.3 任务相关知识

4.2.3.1 PLC 和变频器实现三相异步电动机的启停控制

西门子变频器 G120C PN 的控制单元集成了多种预定义接口宏，用户可以直接调用，从而提高调试效率。本任务采用变频器预定义接口宏 1 实现三相异步电动机的启停控制，即采用两种固定频率的输送技术，变频器相应参数 P0015=1。此方案对应的端子含义如图 4-12 所示。

电机的启停通过数字量输入 DI 0 控制。DI 0 为 1 时电机启动，DI 0 为 0 时电机停止运行。

通过数字量输入选择，可以设置两个固定转速，本任务只设置一个固定转速，数字量输入 DI 4 接通时采用固定转速 3，由 P1003 参数设置。

S7-1215C PLC 的输入由按钮控制，有启动和停止两个按钮。输出 Q0.0 为电机的启动信号，接变频器的端子 5（DI 0）；Q0.1 接变频器的端子 16（DI 4）。当 Q0.0 和 Q0.1 输出均为 1 时，变频器设置为固定转速输出，如电机以 1000r/min 的速度运行。

练一练

(1) 如何实现三相异步电动机的启停控制？将变频器 DI 0、DI 4 相应设置填写到表 4-12 中。

表 4-12 DI 0、DI 4 相应设置

DI 0	DI 4	含义

(2) 根据上面的分析确定 PLC 和变频器端子对应关系，并填写到表 4-13 中。

表 4-13 PLC 和变频器端子对应关系

PLC				变频器		
输入		输出		输入		
功能	元件	地址	地址	端子号	功能	含义
启动						
停止						

4.2.3.2 PLC 和变频器实现三相异步电动机的正反转控制

实现电机正反转控制，可以采用多种方法，本例采用 G120C PN 变频器预定义接口宏 1 实现，即采用两种固定频率的输送技术，变频器相应参数 P0015=1。

电机的正向启停通过数字量输入 DI 0 控制；电机的反向启停通过数字量输入 DI 1 控制。

通过数字量输入选择，可以设置两个固定转速，本任务只设置一个固定转速，数字量

输入 DI 4 接通时采用固定转速 3，由 P1003 参数设置。

S7-1215C PLC 的输入由启动和停止两个按钮控制。输出 Q0.0 为电机的正向启动信号，接变频器的端子 5；输出 Q0.2 为电机的反向启动信号，接变频器的端子 6；Q0.1 接变频器的端子 16。

练一练

（1）如何实现三相异步电动机的正反转控制？将变频器 DI 0、DI 1、DI 4 相应设置填写到表 4-14 中。

表 4-14　DI 0、DI 1、DI 4 相应设置

DI 0	DI 1	DI 4	含义

（2）根据上面的分析确定 PLC 和变频器端子对应关系，并填写到表 4-15 中。

表 4-15　PLC 和变频器端子对应关系

PLC				变频器		
输入		输出		输入		
功能	元件	地址	地址	端子号	功能	含义

4.2.3.3　PLC 和变频器实现三相异步电动机的多段速度控制

实现电机多段速度控制，可以采用多种方法。本例采用 G120C PN 变频器预定义接口宏 1 实现，即采用两种固定频率的输送技术，变频器相应参数 P0015=1。

电机的启动、停止通过数字量输入 DI 0 控制。

通过数字量输入选择，可以设置两个固定转速，数字量输入 DI 4 接通时采用固定转速 3，数字量输入 DI 5 接通时采用固定转速 4。P1003 参数设置固定转速 3，P1004 参数设置固定转速 4。DI 4 与 DI 5 同时接通时采用固定转速 3 + 固定转速 4。

S7-1215C PLC 的输入由按钮控制，有启动和停止两个按钮。输出 Q0.0 为电机的启动信号，接变频器的端子 5（DI 0）；Q0.1 和 Q0.2 分别接变频器的端子 16（DI 4）和 17（DI 5）；当 Q0.0 和 Q0.1 输出均为 1 时，变频器设置为第一段速度输出，如电机以 800 r/min 的速度运行；当 Q0.0 和 Q0.2 输出均为 1 时，变频器设置为第二段速度输出，如电机以 400r/min 的速度运行；当 Q0.0、Q0.1 和 Q0.2 输出全部为 1 时，变频器设置为第三段速度输出，如电机以 1200r/min 的速度运行。

练一练

（1）如何实现三相异步电动机的三段速度控制？将变频器 DI 0、DI 4、DI 5 相应设置填写到表 4-16 中。

表 4-16 DI 0、DI 4、DI 5 相应设置

DI 0	DI 4	DI 5	含义

（2）根据上面的分析确定 PLC 和变频器端子对应关系，并填写至表 4-17 中。

表 4-17 PLC 和变频器端子对应关系

PLC				变频器		
输入			输出	输入		
功能	元件	地址	地址	端子号	功能	含义

4.2.4 任务实施

在 XK-SX2C 高级维修电工实训台上，完成以下 3 个任务：

（1）PLC 和变频器实现三相异步电动机的启停控制、正反转控制和多段速控制。
（2）要求每个任务在确保触摸屏、PLC 和变频器接线均正确的基础上，接通电源。
（3）注意安全意识和团队意识等职业核心素养的养成。

4.2.4.1 PLC 和变频器实现三相异步电动机的启停控制

任务要求

通过 S7-1215C PLC 的输出与 G120C PN 变频器的数字量输入端子连接，实现对电机的启停控制。完成后对本项目进行评分，详见表 4-24。

操作步骤

其操作步骤分别为：
（1）接线。按照图 4-16 完成 PLC 和变频器的接线工作。
（2）变频器参数设置。其操作步骤为：
1）对变频器进行基本调试。
2）按照电机实际参数设置电机参数。如：

① `MOT VOLT P304` OK 额定电压 380V；

② `MOT CURR P305` OK 额定电流 1.12A；

③ `MOT POW P307` OK 额定功率 0.180kW；

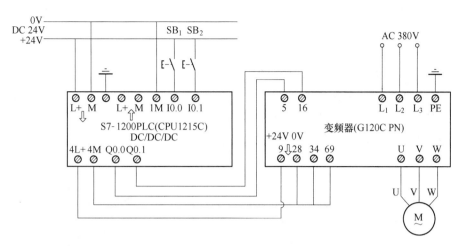

图 4-16 正向启停控制系统接线图

④ MOT FREQ P310　　额定频率 50Hz；

⑤ MOT RPM P311　　额定转速 1430r/min。

3) 选择预定义接口宏。如：

MAc PAr P15　　P0015=1。

4) 电机转速相关设置。如：

① MIN RPM P1080　　电机的最小转速设定为 0；

② MAX RPM P1082　　电机的最大转速设定为 1430；

③ RAMP UP P1120　　电机的加速时间设置为 0.005；

④ RAMP DWN P1121　　电机的减速时间设置为 0.005；

⑤ OFF3 RP P1135　　结束基本调试：使用箭头键切换 NO-YES，按下〈OK〉键。

按照表 4-18 设置变频器参数。

表 4-18　参数设置

参数号	参数值	功　能
P0015	1	预定义接口宏 1
P0840	722.0	将 DI 0 作为启动信号，r722.0 为 DI 0 状态的参数
P1000	3	固定频率运行
P1016	1	固定转速模式采用直接选择方式
P1022	722.4	将 DI 4 作为固定设定值的选择新信号，r722.4 为 DI 4 状态的参数
P1003	1000	定义固定设定值 3，单位为 r/min
P1070	1024	定义固定设定值作为主设定值

(3) PLC。硬件组态是指添加新设备,选择实训台相应的 PLC 型号并设置相应 IP 地址。其步骤如下:

1) 打开 TIA Portal 软件,创建 S7-1200 项目。

2) 打开项目视图,点击"添加新设备",弹出添加新设备对话框,选择控制器,具体路径为:SIMATIC S7-1200->CPU->CPU 1215C DC/DC/DC->6ES7 215-1AG40-0XB0,如图 4-17 所示,添加的 PLC 如图 4-18 所示。

3) 双击图 4-18 中的以太网接口处,设置 PLC 的 IP 地址,设置为 192.168.0.1,如图 4-19 所示。

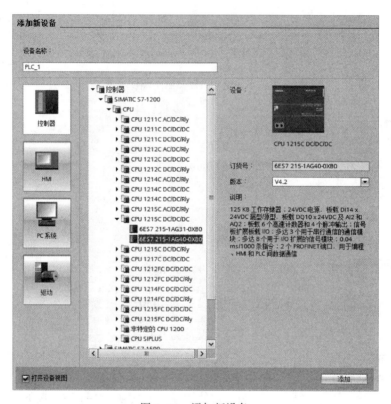

图 4-17 添加新设备

软件编程是指编写程序并下载调试。其步骤如下:

1) 根据图 4-16 确定 I/O 地址分配,填入表 4-19 中。

表 4-19 I/O 地址分配表

输入信号				输出信号		
序号	功能	元件	地址	序号	变频器端子名称	地址

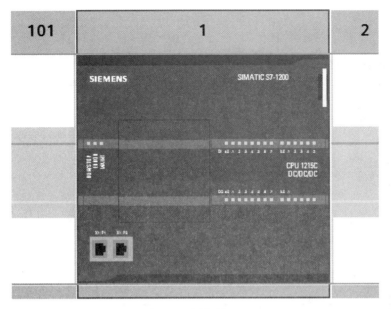

图 4-18 添加的 PLC

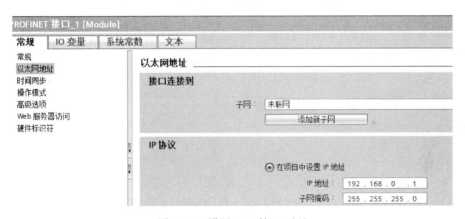

图 4-19 设置 PLC 的 IP 地址

2) 编写程序，下载到 PLC 中进行调试。设备组态后，需要在程序块中编写程序，简单程序可以直接写到 Main [OB1] 中，也可以添加新块，在新块中编写程序，通过 OB1 调用。本例直接在 OB1 中编写程序，双击"Main [OB1]"进入如图 4-20 所示画面。参考程序如图 4-21 所示。

程序编辑完成后，下载到 PLC 中并监控程序，看运行状态是否满足要求。按下启动按钮，监控结果如图 4-22 所示，按下停止按钮，监控结果如图 4-23 所示。

3) 把三相异步电动机接入线路中，通电观察其运行状态。

4) 总结并记录。

分析总结

总结整个任务实施过程，查找不足，尤其对于出现的故障要高度重视，分析总结后记录在表 4-25 中。

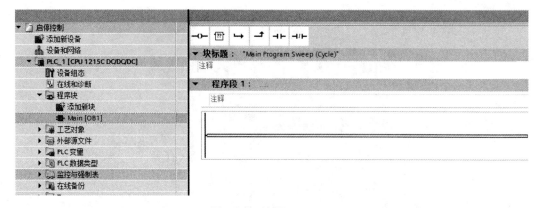

图 4-20　编程入口 OB1

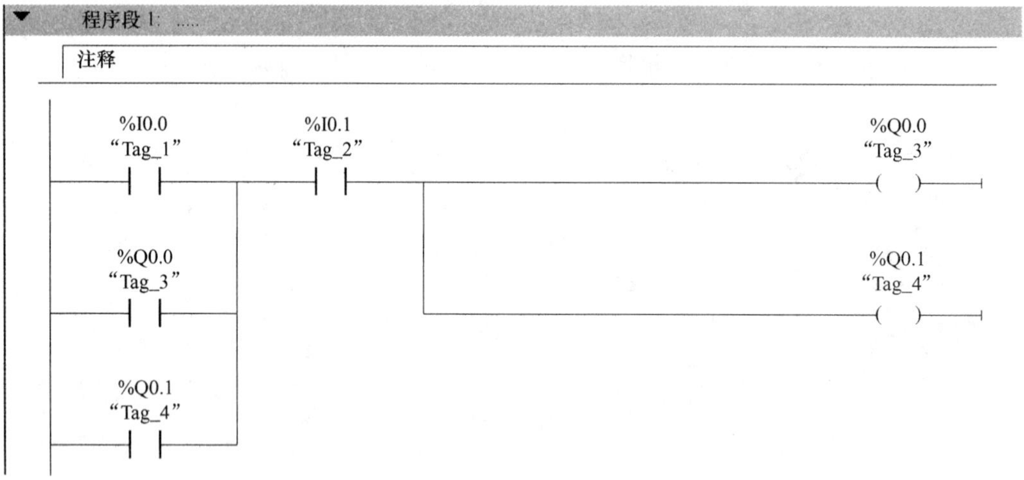

图 4-21　参考程序

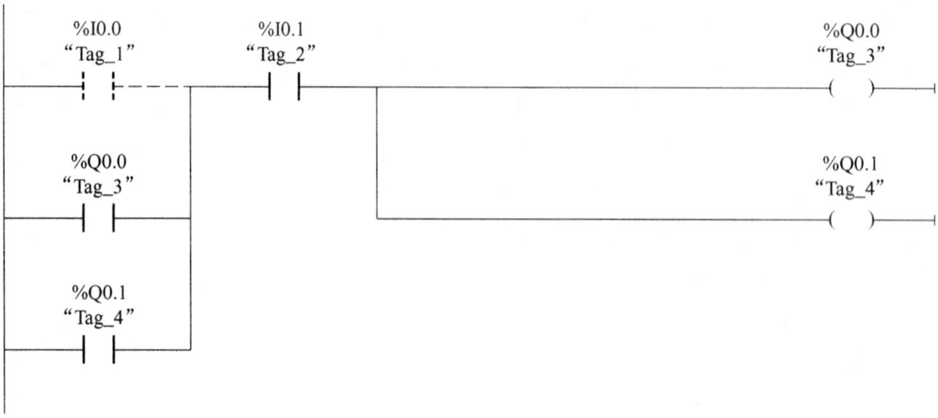

图 4-22　按下启动按钮监控程序结果

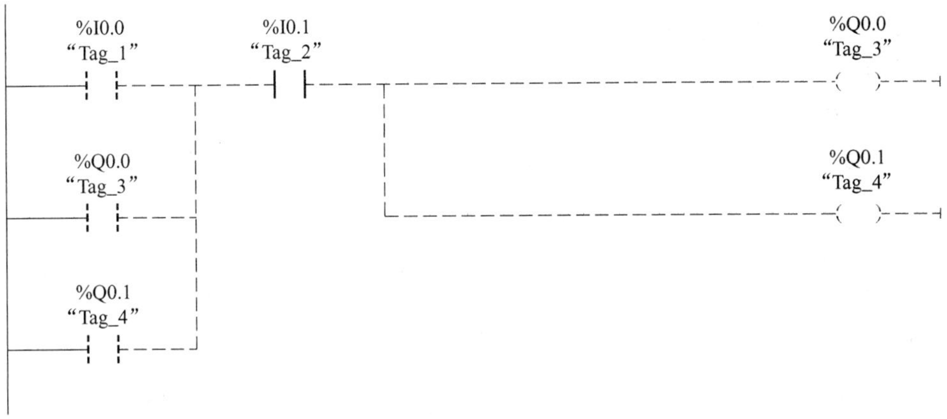

```
  %I0.0        %I0.1                                    %Q0.0
  "Tag_1"      "Tag_2"                                  "Tag_3"
───┤ ├──────────┤ ├──────────────────────────────────────( )───

  %Q0.0                                                 %Q0.1
  "Tag_3"                                               "Tag_4"
───┤ ├─────────────────────────────────────────────────( )───

  %Q0.1
  "Tag_4"
───┤ ├───
```

图 4-23 按下停止按钮监控程序结果

4.2.4.2 PLC 和变频器实现三相异步电动机的正反转控制

任务要求

通过 S7-1215C PLC 的输出与 G120C PN 变频器的数字量输入端子连接，实现对电机的正反转控制。具体控制要求如下：

（1）当按下启动按钮，PLC 的 Q0.0 和 Q0.1 输出均为 1，电机正转，以 1000r/min 的速度运行。

（2）5s 后，PLC 的 Q0.2 和 Q0.1 输出均为 1，电机反转，以 1000r/min 的速度运行。

（3）任何时候按下停止按钮，电机立即停止转动。

完成后对本项目进行评分，详见表 4-24。

操作步骤

其操作步骤分别为：

（1）接线。按照图 4-24 所示，完成 PLC 和变频器的接线工作。

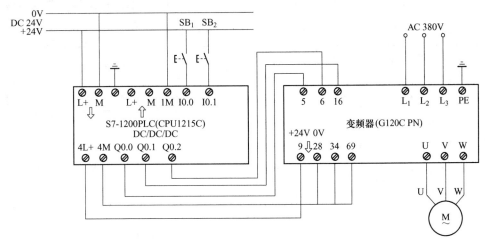

图 4-24 正反转控制系统接线图

（2）变频器参数设置。其操作步骤为：
1）对变频器进行基本调试。
2）按照电机实际参数选择电机参数。
3）选择预定义接口宏。
4）电机转速相关设置。

按照表 4-20 设置变频器参数。

表 4-20 参数设置

参数号	参数值	功 能
P0015	1	预设置 1
P0840	r722.0	将 DI 0 作为启动信号，r722.0 为 DI 0 状态的参数
P1113	r722.1	将 DI 1 作为反转信号，r722.1 为 DI 1 状态的参数
P1000	3	固定频率运行
P1016	1	固定转速模式采用直接选择方式
P1022	722.4	将 DI 4 作为固定设定值的选择新信号，r722.4 为 DI 4 状态的参数
P1003	1000	定义固定设定值 3，单位为 r/min
P1070	1024	定义固定设定值作为主设定值

（3）PLC。硬件组态是指添加新设备，选择实训台相应的 PLC 型号并设置相应 IP 地址。步骤如下：

1）打开 TIA Portal 软件，创建 S7-1200 项目。

2）打开项目视图，点击"添加新设备"，弹出添加新设备对话框，选择 CPU1215 DC/DC/DC（6ES7 215-1AG40-0XB0）。

3）将 PLC 的 IP 地址设置为 192.168.0.1。

软件编程：根据图 4-17 确定 I/O 地址分配，填入表 4-21 中，并按照要求编写程序。

表 4-21 I/O 地址分配表

	输入信号			输出信号		
序号	功能	元件	地址	功能	变频器端子名称	地址

4）编写程序，下载到 PLC 中进行调试。参考程序如图 4-25 所示。

5）把三相异步电动机接入线路中，通电观察其运行状态。

6）总结并记录。

分析总结

总结整个任务实施过程，查找不足，尤其对于出现的故障要高度重视，分析总结后记录在表 4-25 中。

程序段 1: ……

注释

```
    %I0.0        %I0.1                                    %Q0.0
   "Tag_1"      "Tag_2"                                  "Tag_4"
  ———| |————————|/|————————————————————————————————————( S )———
                   │
                   │                                     %Q0.1
                   │                                    "Tag_5"
                   └─────────────────────────────────( S )———
```

程序段 2: ……

注释

```
    %Q0.0        %M0.1                                    %M0.0
   "Tag_4"      "Tag_7"                                  "Tag_3"
  ———| |————————|/|————————————————————————————————————( )———
    │
    │  %M0.0                          %DB1
    │ "Tag_3"                   "IEC_Timer_0_DB"
    └——| |——                       ┌──────────┐
                                   │   TON    │
                                   │   Time   │          %M0.1
                                   │          │         "Tag_7"
                                   ┤IN       Q├────────( )———
                              T#5s ┤PT       ET├ …
                                   └──────────┘
```

程序段 3: ……

注释

```
  "IEC_Timer_0_
     DB".ET                                              %Q0.0
       ==                                               "Tag_4"
      Time   ————————————————————————————————————————( R )———
      T#5s       │
                 │                                      %Q0.2
                 │                                     "Tag_6"
                 └─────────────────────────────────( S )———
```

图 4-25 正反转控制系统参考程序

4.2.4.3 PLC 和变频器实现三相异步电动机的多段速度控制

任务要求

通过 S7-1215C PLC 的输出与 G120C PN 变频器的数字量输入端子连接，实现对电机的多段速度控制。本例第一段速度设为 800r/min，第二段速度设为 400r/min，第三段速度设为 1200r/min。

完成后对本项目进行评分，详见表 4-24。

操作步骤

其操作步骤分别为：

（1）接线。按照图 4-26 所示，完成 PLC 和变频器的接线工作。

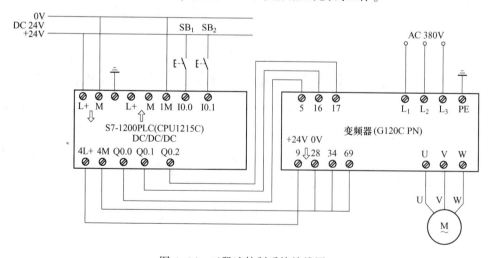

图 4-26 三段速控制系统接线图

· 272 ·

(2) 变频器参数设置。其操作步骤为：

1) 对变频器进行基本调试。
2) 按照电机实际参数选择电机参数。
3) 选择预定义接口宏。
4) 电机转速相关设置。

按照表 4-22 设置变频器参数。

表 4-22 参数设置

参数号	参数值	功　　能
P0015	1	预设置 1
P0840	722.0	将 DI 0 作为启动信号，r722.0 为 DI 0 状态的参数
P1000	3	固定频率运行
P1016	1	固定转速模式采用直接选择方式
P1022	722.4	将 DI 4 作为固定设定值的选择新信号，r722.4 为 DI 4 状态的参数
P1023	722.5	将 DI 5 作为固定设定值的选择新信号，r722.5 为 DI 5 状态的参数
P1003	800	定义固定设定值 3，单位为 r/min
P1004	400	定义固定设定值 4，单位为 r/min
P1070	1024	定义固定设定值作为主设定值

(3) PLC。硬件组态是指添加新设备，选择实训台相应的 PLC 型号并设置相应 IP 地址。步骤如下：

1) 打开 TIA Portal 软件，创建 S7-1200 项目。
2) 打开项目视图，点击"添加新设备"，弹出添加新设备对话框，选择 CPU1215 DC/DC/DC（6ES7 215-1AG40-0XB0）。
3) 将 PLC 的 IP 地址设置为 192.168.0.1。

软件编程：确定 I/O 地址分配，填入表 4-23 中，并按照要求编写程序。

表 4-23 I/O 地址分配表

输入信号				输出信号		
序号	功能	元件	地址	功能	变频器端子名称	地址

4) 编写程序，下载到 PLC 中进行调试。参考程序如图 4-27 所示。

▼ 程序段 1: .....
　注释

```
    %I0.0      %I0.1      %M0.1                    %M0.0
   "Tag_1"   "Tag_2"    "Tag_3"                  "Tag_4"
───┤P├──────┤/├────────┤/├───────┬───────────────( )────────
    %M0.0                         │   %DB3
   "Tag_3"                        │    "T1"
─────┤├──────                     │   ┌─────────┐     %M0.1
                                  │   │   TON   │   "Tag_3"
                                  └───┤   Time  ├──( )────
                                      │ IN    Q │
                                T#20s─┤ PT   ET ├─...
                                      └─────────┘
```

▼ 程序段 2: .....
　注释

```
   %I0.0                                            %Q0.0
  "Tag_1"                                          "Tag_5"
───┤├──────┬────────────────────────────────────────( S )───
           │                                         %Q0.1
           │                                        "Tag_6"
           └────────────────────────────────────────( S )───
```

▼ 程序段 3: .....
　注释

```
   %M0.1                                            %Q0.1
  "Tag_4"                                          "Tag_6"
───┤├──────────────────────────────────────────────( S )────
```

▼ 程序段 4: .....
　注释

```
   %I0.1                                            %Q0.0
  "Tag_2"                                          "Tag_5"
───┤├──────────────────────────────────────────( RESET_BF )─
                                                     3
```

程序段 5: ......

注释

```
    "T1".ET                                              %Q0.2
    ┤ == ├                                              "Tag_8"
     Time                                                ─( S )─
     T#10s
                                                         %Q0.1
                                                        "Tag_6"
                                                         ─( R )─
```

图 4-27　三段速控制系统参考程序

5) 把三相异步电动机接入线路中，通电观察其运行状态；

6) 总结并记录。

分析总结

总结整个任务实施过程，查找不足，尤其对于出现的故障要高度重视，分析总结后记录在表 4-25 中。

4.2.5　任务评价

对第 3 个任务 PLC 和变频器控制三相异步电动机的多段速度控制操作任务的完成情况进行评分，并将评分结果填入表 4-24 中。

表 4-24　评分表

任务内容	考核要求	评分标准	配分	扣分	得分
电路设计	正确设计 PLC 和变频器控制电机运行线路接线图	(1) 电气控制原理设计功能不全，每缺一项功能扣 5 分； (2) 接线图表达不正确或画图不规范每处扣 2 分	10		
电路接线	按照电路图正确完成接线	(1) 每接错一根线扣 5 分； (2) 其他情况酌情扣分	10		
变频器参数设置与调试	按照任务规定的控制要求进行变频器相应参数的设置	(1) 变频器参数设置不全，每项扣 3 分； (2) 参数设置错误每处扣 2 分； (3) 不会设置参数扣 20 分	20		

续表 4-24

任务内容	考核要求	评分标准	配分	扣分	得分
PLC 程序编写与下载	按照任务要求将 PLC 程序输入、下载与监控	（1）软件使用不熟练，不能完成程序的输入操作扣 15 分； （2）不会进行程序的 IP 地址设置与下载扣 5 分； （3）不会操作博途软件监控程序的运行扣 5 分	20		
PLC 控制系统运行演示	正确演示电动机的启动和停止，并能结合程序、硬件进行说明	（1）电路不能实现启动扣 10 分； （2）电路不能实现停止扣 10 分； （3）功能实现错误或不全扣 10 分； （4）不能对操作现象进行分析和说明的每处扣 5 分	30		
安全文明生产	遵守 8S 管理制度，遵守安全管理制度	（1）未穿戴劳动保护用品扣 10 分； （2）操作存在安全隐患，每次扣 5 分，扣完为止； （3）操作现场未及时整理整顿，每次发现扣 2 分，扣完为止	10		
总 分					

4.2.6 任务小结

任务分析总结记录单见表 4-25。

表 4-25 任务分析总结记录单

	任务分析总结记录单	
故障 1	故障现象	
	故障原因	
	排除过程	
故障 2	故障现象	
	故障原因	
	排除过程	
故障 3	故障现象	
	故障原因	
	排除过程	
总 结		

4.2.7 任务拓展

任务要求：通过 S7-1215C PLC 的输出与 G120C 变频器的数字量输入端子连接，实现对电机的正反多段速度控制。三段速度设为 800r/min、400r/min 和 1200r/min。具体控制要求如下：

(1) 当按下启动按钮，PLC 的 Q0.0 和 Q0.1 输出均为 1，电机正转，以 800r/min 的速度运行。

(2) 5s 后，PLC 的 Q0.0 和 Q0.2 输出均为 1，电机正转，以 400r/min 的速度运行。

(3) 10s 后，PLC 的 Q0.0、Q0.1 和 Q0.2 输出全部为 1，电机正转，以 1200r/min 的速度运行。

(4) 15s 后，PLC 的 Q0.3 和 Q0.1 输出均为 1，电机反转，以 800r/min 的速度运行。

(5) 20s 后，PLC 的 Q0.3 和 Q0.2 输出均为 1，电机反转，以 400r/min 的速度运行。

(6) 25s 后，PLC 的 Q0.3、Q0.1 和 Q0.2 输出全部为 1，电机反转，以 1200r/min 的速度运行。

(7) 30s 后，PLC 输出为 0，电机停止转动。

任何时候按下急停按钮，电机立即停止转动。选择预定义接口宏 1，接线图如图 4-28 所示。

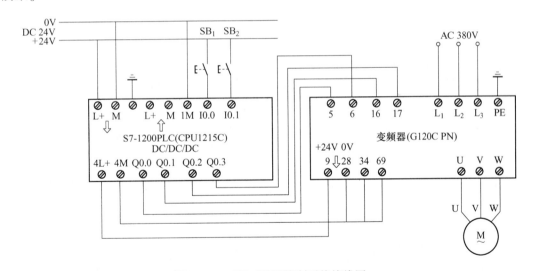

图 4-28 正反三段速控制系统接线图

4.2.8 项目小结

通过任务 4.1 的实施，学习了变频器的基本结构组成与工作原理，并以 SINAMICS G120C 变频器为例，在掌握变频器的端子功能、接线及 BOP-2 基本操作面板的基础上，通过 5 个操作任务的训练，掌握变频器的基本操作。

通过任务 4.2 的学习和实践，掌握了 PLC 和变频器控制三相异步电动机的启停、正反转和多段速度运行的控制。通过 3 个操作任务的训练，掌握了 PLC 对西门子 G120 变频器控制的接线方法；利用 G120C PN 变频器的预定义接口宏 1 初步探索了宏的功能，为应用其他宏功能提供了基础；初步掌握了 PLC 和变频器配合使用控制三相异步电动机的基本操作。

练习题

（1）简述变频器的结构组成及工作原理。

（2）变频器在运行过程中，如何改变它的转向？

（3）如何实现电动机的点动控制？如何修改点动频率？

（4）G120C 变频器的数字量输入、输出端子有哪些？

（5）G120C 变频器的模拟量输入、输出端子有哪些？

（6）变频器有哪些控制方式？

（7）除了预设置方案 1，还有哪些预设置方案能实现三相异步电动机的启停、正反转和多段速度运行的控制？

（8）如何实现模拟量控制的变频调速？

（9）如何实现变频器的无级调速？

参考文献

[1] 陈宝玲. 电气控制技术［M］. 大连：大连理工大学出版社，2019.
[2] 西门子公司. 博途 V14 软件使用手册，2017.
[3] 靳哲. 可编程序控制器原理及应用［M］. 北京：北京师范大学出版社，2010.
[4] 陈建明. 电气控制与 PLC 应用：基于 S7-1200 PLC［M］. 北京：电子工业出版社，2020.
[5] 向晓汉. 西门子 S7-1200 PLC 学习手册：基于 LAD 和 SCL 编程［M］. 北京：化学工业出版社，2018.
[6] 吴繁红. 西门子 S7-1200 PLC 应用技术［M］. 北京：电子工业出版社，2017.
[7] 段礼才，等. 西门子 S7-1200 PLC 编程及使用指南［M］. 北京：机械工业出版社，2018.
[8] 哈立德. 卡梅尔，艾曼. 卡梅尔. PLC 工业控制［M］. 北京：机械工业出版社，2018.
[9] 侍寿永. S7-300PLC、变频器与触摸屏综合应用教程［M］. 北京：机械工业出版社，2018.
[10] 张忠权. SINANICS G120 变频器控制系统使用手册［M］. 北京：机械工业出版社，2016.
[11] 西门子公司. SINAMICS G120C 变频器精简版操作说明，2016.
[12] 马宏骞，许连阁，石敬波，等. PLC、变频器与触摸屏技术及实践［M］. 北京：电子工业出版社，2016.